# Praise for Bernd Heinrich and *Summer World*

"It may not be every urbanite's idea of a dream date, but mine, after reading *Summer World*, is to spend a summer day with a sixty-nine-year-old insect physiologist and all the tools of his trade. . . . Our date would start before dawn and would include, but not be limited to, climbing into treetops, slogging through wetlands, and sitting quietly for hours with pencil and notebook, the better to observe and record. An emeritus professor of biology at the University of Vermont, Bernd Heinrich—the object of my admiration—has been doing all this, and writing about it with brio, for decades. Perhaps his most attractive quality, for this reader at least, is his ability to find something intellectually stimulating whenever he steps out the door. . . . The entomologist himself is a Dumbledore of the forest—magical himself for his ability to conjure a riot of life from what others less attuned might consider your standard Northern woodlot. . . . Animals come to life in gripping detail . . . and so does Heinrich as he bounds between his experiments. The man is irrepressible."

—*New York Times Book Review*

"Bernd Heinrich's books open my eyes and help me see the wonder of the natural world. . . . I love the fascinating details of his drawings, the lyricism of his observations, the way he unveils not only the physical workings of nature but the stories and dramas within it."

—Amy Tan, bestselling author of
*The Bonesetter's Daughter* and *The Joy Luck Club*

"This lovely book, meticulously etched and based on impassioned but exacting scientific research, illustrates why Bernd Heinrich is generally regarded as the most truly Thoreauvian of modern natural history writers."

—Edward O. Wilson,
Pulitzer Prize–winning author of *On Human Nature*

"Heinrich embarks on a project of paying attention, in this case, to a clearing in the woods outside his cabin in Maine as it comes to life repeatedly over the course of several summers. This is hands-and-knees science at its most engaging. Heinrich is an experimenter and a meddler. . . . Anyone who wants to feel more engaged with the trees, bugs, and birds of summer should enjoy *Summer World*."

—Anthony Doerr, *Boston Globe*

"When warm spring days turn blustery, it's useful to have a book writhing with the magic of summer. Bernd Heinrich delivers. . . . In *Summer World*, the longtime author and biology professor shares reflections and experiments from his homes in Vermont and Maine. His illustrations are beautiful . . . and his thoughts are often educational."

—*Seattle Times*

"Bernd Heinrich is one of our greatest living naturalists in the tradition of Gerald Durrell; he's John Muir (without the wandering), Edward Abbey (without the politics), Jacques Cousteau (without the ocean), Ernest Seton (without the talking animals). Heinrich, author of fifteen marvelous, mind-altering books . . . is a national treasure."

—*Los Angeles Times*

"Arguably today's finest naturalist author. . . . Our latter-day Thoreau."

—*Publishers Weekly*

"Heinrich is as brilliant at depicting the highs and lows of scientific research as he is in sharing the ways and wonders of the natural world. And always, always, there is in Heinrich's every page, wonderment."

—*San Francisco Chronicle*

# Summer World

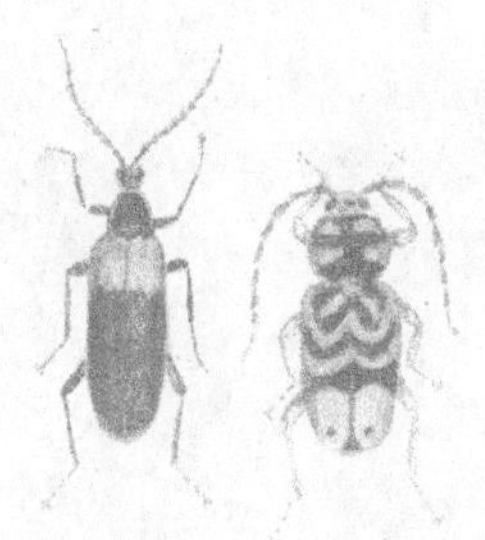

## Also by Bernd Heinrich

*The Snoring Bird*
*The Geese of Beaver Bog*
*Why We Run (previously titled Racing the Antelope)*
*Winter World*
*Mind of the Raven*
*Bumblebee Economics*
*The Trees in My Forest*
*One Man's Owl*
*Ravens in Winter*
*A Year in the Maine Woods*
*The Hot-Blooded Insects*
*In a Patch of Fireweed*
*The Thermal Warriors*

Bernd Heinrich

# Summer World

## A Season of Bounty

ecco

An Imprint of HarperCollins*Publishers*

 For information, address HarperCollins Publishers, 195 Broadway, New York, NY 10007.

HarperCollins books may be purchased for educational, business, or sales promotional use. For information, please e-mail the Special Markets Department at SPsales@harpercollins.com.

A hardcover edition of this book was published in 2009 by Ecco, an imprint of HarperCollins Publishers.



FIRST ECCO PAPERBACK EDITION PUBLISHED 2010.

All line drawings are by Bernd Heinrich.

*Designed by Shubhani Sarkar*

Library of Congress Cataloging-in-Publication Data is available upon request.

ISBN: 978-0-06-074218-8

HB 06.27.2022

To Rachel

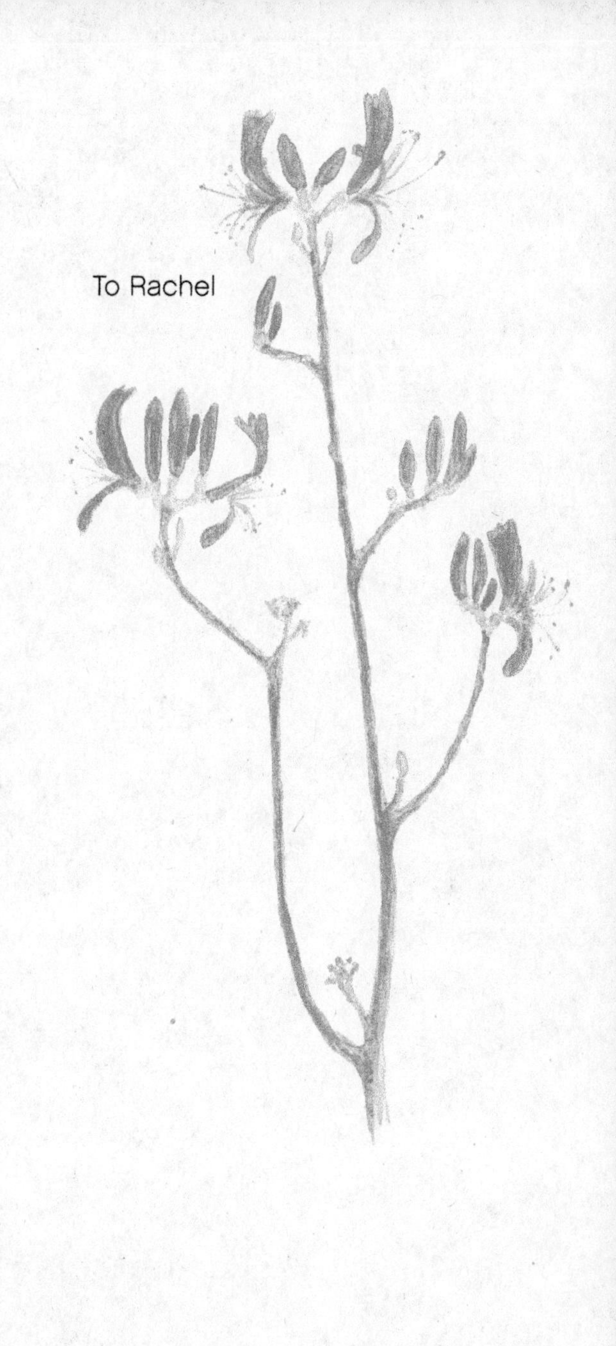

# Acknowledgments

WRITING THIS BOOK REQUIRED ME TO TAP INTO EXPERT advice on many subjects. I leaned on numerous people for help and received advice, suggestions, and inspiration. In alphabetical order, I thank Jeffrey Boettner, Chris Bouchard, Michael Caduto, Janice Cahill, Alice Calaprice, Michael Canfield, Rod Eastwood, Jeffrey Hughes, Daniel Janzen, William Jordan, Frank J. Joyce, Steven Krauth, Kevin O'Neil, Catherine Paris, Naomi Pierce, David L. Wagner, and Edward O. Wilson. I thank Paula Kelly and Lisa Barrett for steering me to hornet nests. But mostly I thank Daniel Halpern, publisher of Ecco, for suggesting this project as a companion and counterpart to *Winter World*; Sandra Dijkstra, my agent, for supporting it; Taryn Fagerness, John Scibetta, and Trey Shores for being a sounding board on several chapters; and Emily Takoudes for her insightful and wise editing, which helped bring it all to fruition.

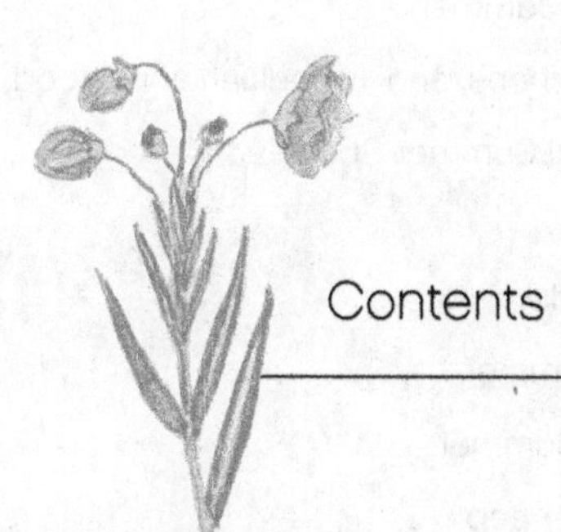

# Contents

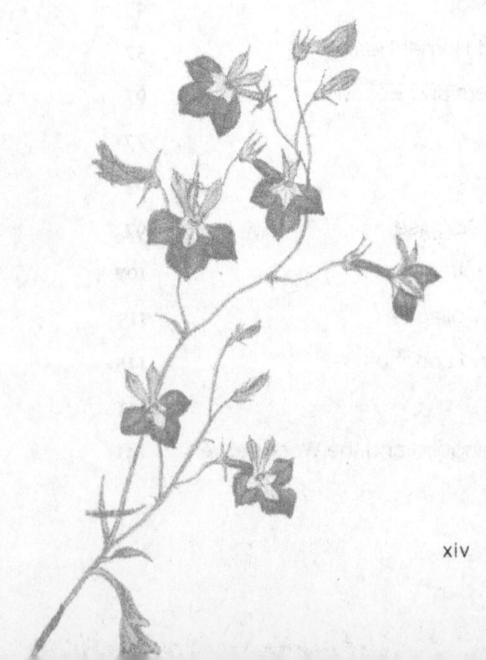

# Summer World

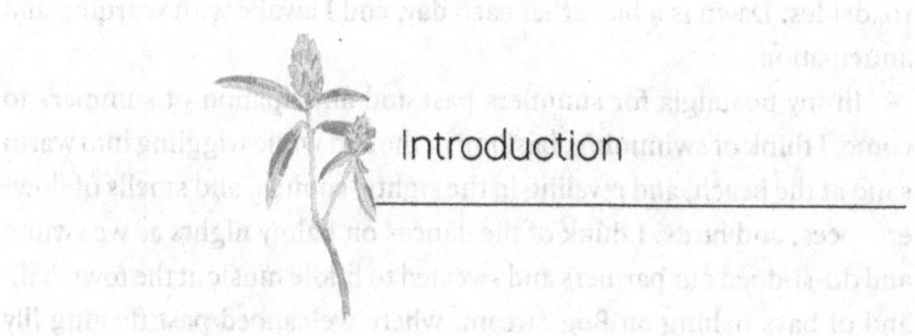

# Introduction

March often brings heavy snowfalls here in Maine and Vermont. It's cold outside and I spend much of my time behind windowpanes in a bubble of tropical environment created by our wood-burning stove. I'm waiting for summer. Here in the north temperate zone, "summer" usually lasts for roughly half the year, from May through October. It's the time most of us (who are not recreational skiers) live or at least wait for.

Day after day I gaze at the white expanse of the beaver bog by our house to wait and hope for the red-winged blackbirds to return. Instead, in my mind's eye during March I see a family of beavers entombed in their lodge, which sticks up like a big lump above the thick snow-covered ice on the pond. The beavers' bubble for sustainable life is now a mere platform of sticks inches above the ice-cold water. It's barely large enough to move around in, and they live there in continuous darkness. Occasionally one or another of the beaver family holds its breath for several minutes as it dives into the hole next to its platform that is kept ice-free, to bring back a twig and chew off the bark. I identify with these beavers, because like theirs, much of my living is, for months, repressed and in my own bubble. Summer releases it.

The world right now seems dead, but some birds are already

stirring. Hairy and downy woodpeckers have started to drum; black-capped chickadees sound off their "dee-dahs" at dawn; and the first robins have returned, and they hop where snow has melted along roadsides. Dawn is a bit earlier each day, and I awake with yearning and anticipation.

In my nostalgia for summers past and anticipation of summers to come, I think of swimming, basking in the sun while wiggling into warm sand at the beach, and reveling in the sights, sounds, and smells of flowers, bees, and birds. I think of the dances on balmy nights as we swung and do-si-doed our partners and sweated to fiddle music at the town hall; and of bass fishing on Bog Stream, where we canoed past floating lily pads and big white water lily blossoms. I think of the school year coming to a close.

For me, summer used to begin on the first day of school vacation, the season of long days. A more universal and just as specific beginning of summer (in the northern hemisphere) is probably around 20 March, the vernal (spring) equinox ("equal night"), when the night and the day are the same length. The height of northern summer is near 21 June, the summer solstice (corresponding to the winter solstice in the southern hemisphere), when in the north the days are the longest and we receive the most sunshine in the year. However, this is designated as the beginning of summer, not the height, because the maximum warmth is yet to come; it takes about a month and a half before the northern lands and oceans, still cold from the winter, have reheated. Then, after the summer solstice, the days shorten until about ninety-four days later, on 22 September, when they are again equal. On 21 December, the winter solstice, the days are shortest. Again, owing to the temperature lag from the just-cooled earth and ocean, this date is called the beginning of winter, not its peak.

Almost all of life on the surface of the earth is fueled by the enormous amounts of energy intercepted from the sun, through a chemical reaction involving one main molecule, chlorophyll, and its reaction with water and carbon dioxide to produce sugar, the main fuel that powers life. The process that produces it is photosynthesis, meaning, literally, "making from photons." The amount of this energy that continually streams onto Earth, and is proximally fixed into sugar, is relatively constant throughout the year, but the portion that is captured in any one place on Earth at any one time depends largely on the daily duration of illumination, and the angle at which the rays hit the Earth's surface.

Both the duration and the incidence of illumination at any one place depend on the Earth's tilt, or inclination, toward the sun, and the seasons are a consequence of this tilt. At all points of the Earth's approximately 365-day (actually 365.2422-day) orbit around the sun, which we define as a year, the Earth's axis of rotation (an imaginary line connecting the north and south poles) is 23.5 degrees with respect to the plane of its orbit around the sun. This angle does not affect the total energy that the entire Earth receives over the year; rather, it shifts the distribution of energy between the northern and southern hemispheres. When one hemisphere gets a lot of energy, the other gets little, and thus when it is summer in one it is winter in the other. At the equator the energy input is equal year-round, the sun is directly overhead at noon, and days and nights are always equal.

When the Earth is at the point in its orbit where the north pole is inclined at its maximum, 23.5 degrees, toward the sun, that is defined as the summer solstice in the north. At this time the far north is in continuous light and the far south is in continuous darkness. As the Earth continues its journey around the sun (while still maintaining its own same axis of rotation) the tilt that was toward the sun decreases gradually until solar radiation falls equally slanted onto both poles. At this point, the autumnal equinox, day and night are of equal length everywhere.

The solstices, the asteorological relationships during the Earth's annual journey around the sun, proximally cause the seasons and the overall weather patterns to which life adjusts. However, ultimately the seasons are due to an ancient catastrophe. Astronomers believe that about 4 billion years ago a body the size and mass of Mars slammed into the Earth at 18,000 miles per hour, possibly tipping the Earth's axis of rotation. Additionally, the matter that was ejected by this colossal collision produced the moon. Life arose near that time, about half a billion (500 million) years later, and it has adjusted to summer versus winter ever since. Different species each have their own schedules of preparation for summer, although for most summer is the season of reproduction, feeding, growing, and trying to avoid being eaten. It's the season of courting, mating, and birthing; of living and dying.

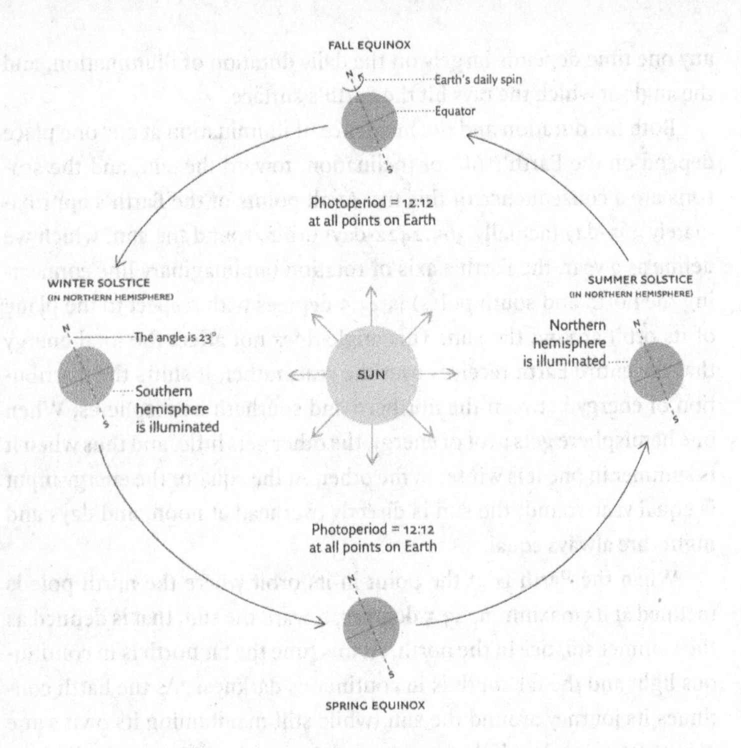


Fig. 1. **The Earth's annual journey around the sun, showing the seasons in relation to the planet's tilt. At the solstices, the durations of day and night are different in the north versus the south polar regions. At the equinoxes these durations are equal at every point on Earth.**

---

THE MEMBERS OF MANY SPECIES, MYSELF INCLUDED, become more alive again at the first "scent" of summer. Skunks come out of their dens, and we get the first whiffs of their presence. Chipmunks emerge from underground and leave their first tracks on the softening snow. The yearling beavers leave their dens as their parents get ready to have new pups. Flower buds of the willows, alder, beaked hazel, poplars, and elms are poised to respond to the first warmth, to open and reveal their beautiful colors and varied forms. Some of the birds that overwintered begin to sing, and the migrant birds are plying the skies by the millions on their way north from the tropics. The first are starting to arrive.

Nature is about to burst at the seams. As George Harrison's song, performed by the Beatles in 1969, goes: "Here comes the sun—da da da da. It's all right—." The warming sun signals relief, and I'm ready. The rest of nature has been waiting and getting ready as well.

During the increasingly longer and brighter days after the vernal equinox, the purple-brown flower buds of the alders in the bog, and those of the birches, hazels, and quaking aspen that surround it, begin to get ready for summer. These plants had their flower buds fully formed in the fall, ready to pop open and bloom at the right moment. Some had already formed their new leaf buds in early July, during the warmth of the previous summer, to get a jump after winter for the brief summer to come. Not all northern buds "hold back" from July till next June. Some "jump the gun"—red oak buds, for example, on shoots that have access to direct sunlight, often "break" in July and produce a second shoot with another set of leaves, instead of waiting eleven months for the next year. Then, however, they still make another set of buds before winter.

To the bees in my two beehives under snow next to our house, the external world will scarcely have changed over the last several months, but they have also been getting ready. The queen will already have started laying eggs into the combs so the hive can field a large cohort of workers to exploit the big but brief first flush of bloom of the poplars and maples, long before the leaves appear.

Summer is "those lazy, hazy, crazy days" that Nat King Cole sang about. But is it more? I asked my eight-year-old daughter, Lena, to tell me what she thought it is, and she wrote me a poem that I give here verbatim: "Summer days are fun. They make me want to run, under the hot boiling sun! The days are long and light, I get to stay up late at night! It's quite a starry sight! Screaming yelling, chanting! Running, jogging, panting!" I wonder where she gets her ideas, but to me her poem seems in tune with Roger Miller's catchy rhythm and words from the 1960s: "In the summertime, when all the trees and leaves are green and the redbird sings, I'll be blue, 'cause you don't want my love."

Summer is a time of green, urgency, and lots of love lost and found. It is the most intense time of the year, when the natural world of the northern hemisphere is almost suddenly populated with billions of animals awakening from dormancy, and billions more arriving from the tropics. Almost overnight there is a wild orgy of courting, mating, and rearing young. The main order of business in summer is reproduction, and the window of opportunity is short. Proximally, summer may be a frolic, but that masks the underlying competition and struggle, because for every new life of any one species there are necessarily, on average, equal numbers of deaths of that same species. Furthermore, for each of the large animals there are necessarily also hundreds or thousands of deaths of smaller ones of other species that get eaten to produce this life. And every one of them has evolved mechanisms to reduce its chances of being eaten.

THE KEY TO SURVIVAL IN WINTER IS FINDING SOLUTIONS to a combination of cold and scarce energy. Summer is the opposite situation. One could consider the summer world as delineated by survival at high temperatures and limited by water; and although I provide a brief reference to the "extreme" summer of physical constraints, as in the desert, I have chosen instead to look at the ingenuity of life more locally, as life-forms interact with one another—the main order of business in the summer. I focus mainly on what I see and saw played out in the familiar world that is at the doorstep of my log cabin in Maine, in a clearing in the woods. In addition, I paid as much or more attention to all this at our home along a dirt road in rural Vermont. Our house is surrounded by woods, a beaver bog, a vegetable garden, a couple of beehives, bird boxes, a woodshed, and patches of wild and cultivated flowers and fruit trees. I decided to live two summers actively observant. I wanted to pursue the interesting and often puzzling, without taking the seemingly prosaic for granted.

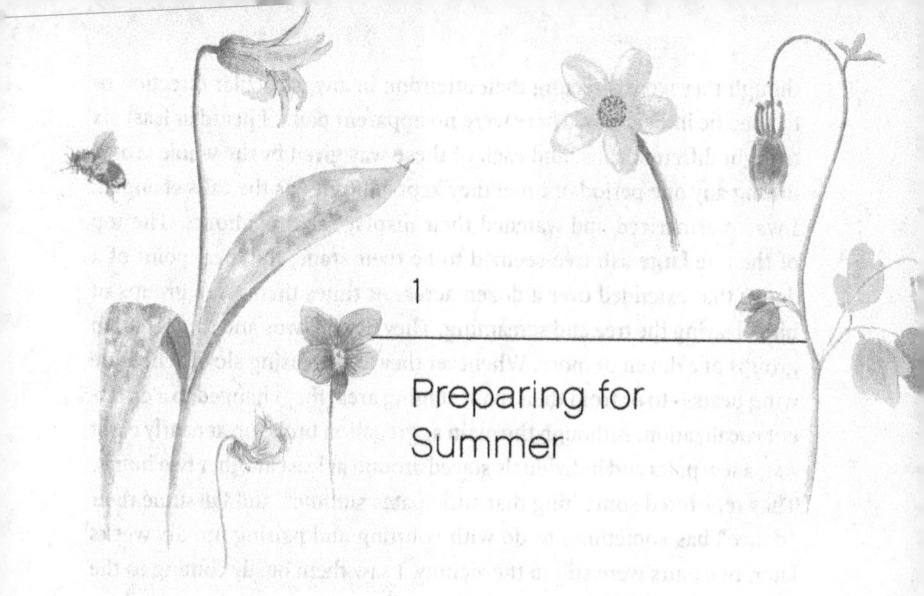

# 1
# Preparing for Summer

9 March 2006. THE GROUND IS STILL SNOW-COVERED, but I've smelled the first skunk, and the bog is threaded by mink and otter tracks. I've heard the first honking of Canada geese. Two big flocks flew over, very high, heading north. The plant life looks unchanged, except that some pussy-willow buds have recently started to show a little more white peeking out over the edges of their dark brown flower bud scales. The first snowdrops, in the pure, unassuming simplicity that I love, are poking their nodding flower heads through the snow. Yesterday evening I heard the first singing of a mourning dove. The first robin is back, long before a worm is in sight. It's overcast and the forecast says "rain," but even if snow were predicted I'd expect the male red-winged blackbirds to return any day now.

Spring is on the way, and I think the birds feel it too. Certainly the blue jays do. I was lucky to see their first convocation again this year. I first noticed a crowd of them making a racket at seven AM on the top of the bare branches of an ash tree—the same one where I saw them about this time last year. I counted at least twenty-four, but these jays were coming and going, so maybe there were many more. Those in the top of the tree were bobbing up and down in what looked like energetic knee-bend exercises, and calling at the same time. It did not look as though they were directing their attention in any particular direction or to specific individuals. There were no apparent pairs. I heard at least six to eight different calls, and each of these was given by the whole crowd during any one period of time; they kept "in tune" as the calls changed. I was mesmerized and watched their display for three hours. The top of the one large ash tree seemed to be their stage, the focal point of a dance that extended over a dozen acres. At times there were groups of birds leaving the tree and screaming. They flew in twos and threes and in groups of a dozen or more. Whenever they went—using slow, deliberate wing beats—to or from their main staging area, they changed to a different vocalization. Although the main aggregation broke up at nearly eight AM, a few pairs and individuals stayed around at least another two hours. They registered something that anticipates summer, and I assume their "dance" has something to do with courting and pairing up. Six weeks later, two pairs were still in the vicinity. I saw them busily coming to the edge of my recently dug frog pond to pull rootlets out of the ground for lining their nests.

SUMMER IN THE NORTHERN HEMISPHERE IS SHORT, AND preparation for it is long. Getting an early start in the race to reproduce is critical. The season is anticipated by most organisms through photoperiod, the relative hours of day versus night. The seasons can also potentially be read from the stars. During summer, fall, winter, and spring in the northern hemisphere the North or "Pole" Star, Polaris, is visible as a steady fixed point above the horizon. Its angle to the Earth used to indicate latitude to the mariner. Daily, the constellations turn once around this star, rising in the east and setting in the west. Nearest to it we see the Big Dipper and Little Dipper and Cassiopeia. All three constellations are visible throughout the year, although in winter, when the tilt of the Earth's northern hemisphere is away from the sun, a new direction of the sky that was blocked during summer comes into view, with other constellations. Now the constellation Orion rises on the eastern horizon in the evening and dominates the southern sky, along with Sirius, a large star. During the summer in the northern hemisphere these winter stars are below the horizon, and the overhead sky is dominated by the Milky Way and three brilliant stars: Vega, Deneb, and Altair, of the constellations Lyra, Cygnus, and Aquila. Together these three stars, the "summer triangle," are a clear sign of summer. Do the birds know this?

Whether or not any animals can read and interpret the changing seasons from star patterns, and from them anticipate and prepare for the seasons, is not known. We do know, though, that animals use star patterns for navigation during their migrations. Many birds migrate at night, mainly the small songbirds whose rate of energy expenditure is so great that they need the day to refuel more often than large birds do. They watch the stars and recognize the patterns of the night sky. We know from detailed experiments and observations that they orient to the North Star or, more likely (as we do), by the star patterns around it, such as the Big Dipper. On their northward migrations to their summer breeding grounds they fly toward the Big Dipper, just as slaves escaping north at night followed it, code-naming it the "Drinking Gourd." When the birds return at the end of the summer, Polaris and the Big Dipper are on their backs or shoulders as they fly through the night sky.

Proximally, summer in the northern hemisphere is best defined, as already mentioned, by the period of sunlight and warmth that sustains active life. In the tropics "summer" is essentially endless; there are about 4,320 hours of daylight per year. Here in New England daylight is restricted to about 2,520 hours. And despite the much longer days in the arctic summer, there are fewer of them—not quite half those in New England. However, my calculation is an approximation only. I have for simplicity assumed (1) thirty-day months, (2) twelve months of summer with twelve hours of daylight per day in the tropics, (3) six months of summer with an average of fourteen hours of daylight per day in the temperate zone, and (4) two months of summer with twenty-four hours of light per day in the high arctic.

In early February the worst of the winter is yet to come, even though the days lengthen. On some days when the sun does come out, I hear chickadees singing, blue jays carousing, the great horned owl hooting, and woodpeckers drumming. But the weather, like these activities that depend on it, is unpredictable. In 2006 the spring was unseasonably cold and the fall was unseasonably warm. More snow fell in Vermont that April than had been recorded for more than 100 years. But in early February a raven pair that I know had already refurbished their nest, and the female was incubating a clutch of eggs. A pair of great horned owls then evicted the ravens and took over the nest, and in early April the owl perched on her eggs (or her small young, or both) in the nest mold, which became surrounded by a wall of snow a foot high. The ravens did not renest that year—there wasn't enough time. They, like the owl, have a narrow window of time. They need to have their young independent by fall. So they get an early start on summer. They need all summer and then some. Nest building and incubation take at least a month; rearing the young takes another two months; and then the young adults need the summer to practice their hunting skills, while there are still numerous young animals around to catch.

The trees prepare for the coming summer nine months in advance, starting in July of the previous year, when they manufacture embryonic shoots, leaves, and flowers and enclose them in buds. They could poten-

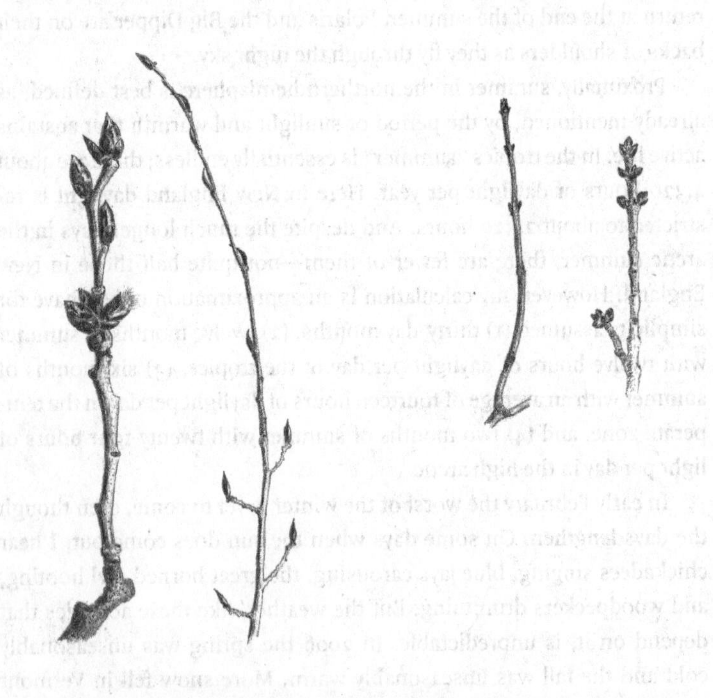

Fig. 2. Leaf and flower buds (in Vermont) of quaking aspen *(left)* and red maple *(right)*. In each pair of twigs the thinner one bears leaf buds and the thicker one (from the top of the tree) bears flower buds.

tially wait until spring (some, like black locusts, which flower late, do), but for the northern native trees it is apparently better to have at least the shoot-leaf buds ready-made to burst forth at a signal. It is too cold to make them in the winter, and their final signal for the buds to burst forth is warmth. The hitch, though, is that they risk death if they are fooled by any warm spell, such as the usual January thaw. Insects also prepare to be active at specific times in the coming summer. For example, the giant silk moths (Saturniidae) overwinter in the pupal stage, and like tree buds, they shut down their development from pupa to adult through late summer, fall, winter, and spring.

Insect development from the pupal into the adult stage is normally strictly temperature-dependent: the higher the temperature, the faster they become adult. But the overwintering moth pupae can hold back even if they experience warmth. Their rather amazing block to the developmental process can be removed only by subjecting them to both a sufficient length and a sufficient depth of cold. As shown by brain transplants, the developmental blockage originates in the insects' brain; implanting a chilled "loose" or unconnected brain that has been subjected to the right day length (depending on the species) into the abdomen of a non-chilled pupa will start the process of development as it releases hormones into its host's blood. Ultimately the pupae, by not activating their normal brain, are preparing and waiting for the next summer, which is about ten months in the future. And only then, at the right time, does the whole population of millions of them emerge relatively synchronously in a week or two to mate and lay their eggs. They must be timely, and quick; they live for only about a week.

Preparing for summer means being able to anticipate the upcoming season, which presupposes knowing (no consciousness is implied) what season you are in. Perhaps one of the most reliable seasonal cues is photoperiod, the relative hours of daylight versus dark in a twenty-four-hour cycle. Throughout late summer and fall the days get shorter, then lengthen again after the winter solstice. Thus, an organism that sees neither the stars nor the angle of the sun could potentially anticipate the approaching summer by registering day lengths.

To measure day length requires the use of a clock that, on our planet, runs on a twenty-four-hour cycle or period. Biological clock mechanisms with approximately this period have by now been demonstrated in one-celled organisms, plants, insects, birds, and mammals. But a clock, even if it has a correct period, is not sufficient to tell time, any more than a watch that has not been set to the local time. Biological clocks must also be set to the correct local time; to do its job, each clock must be sensitive to and synchronized by signals from the environment, in the same way that we set our watches to the time we may hear announced on the radio, or from some other cue.

Like any good watch, a biological clock doesn't run faster or slower as temperatures increase or decrease, even though the individual chemical reactions that run it presumably do. However, like the windup wristwatches that we used to wear (before batteries), which would run perhaps a minute or two either fast or slow, so that we had to reset them every few days, biological clocks are never totally accurate and also need frequent resetting to the local solar time. For example, a circadian clock that runs fifteen minutes fast per twenty-four-hour day would be off by an hour within four days. But what are the clocks set to? Most biological clocks are set at the signal of either lights-out or lights-on, which in nature would normally be dusk and dawn, respectively. They thus indicate the actual time relatively accurately, despite the fact that their periods may not be exactly twenty-four hours. Once a clock is set and running, appropriately timed behaviors can be "read" from it and will be close to local time.

One of the first to show that the twenty-four-hour clock could be used by an animal to synchronize to the season was Erwin Bünning, in studies with the common white cabbage butterfly, *Pieris brassicae*. In the summer the caterpillars of this butterfly proceed without pause from the pupa to the adult stage in a couple of weeks, with the exact duration depending on temperature. In the fall the caterpillars still grow normally; but after they have entered the pupal stage, they stop further development regardless of temperature. If they didn't stop, they would all hatch out when there is no cabbage for their caterpillars to feed on. So they do not continue developing into adults until the following summer. Bünning asked how the animals "know" what season they are in, and what they do about it. He found out that the caterpillars have a clever mechanism involving the use of their daily or twenty-four-hour clock.

Using their circadian clock, the cabbage butterfly larvae begin measuring time at a specific signal: as in most other species, the time of day

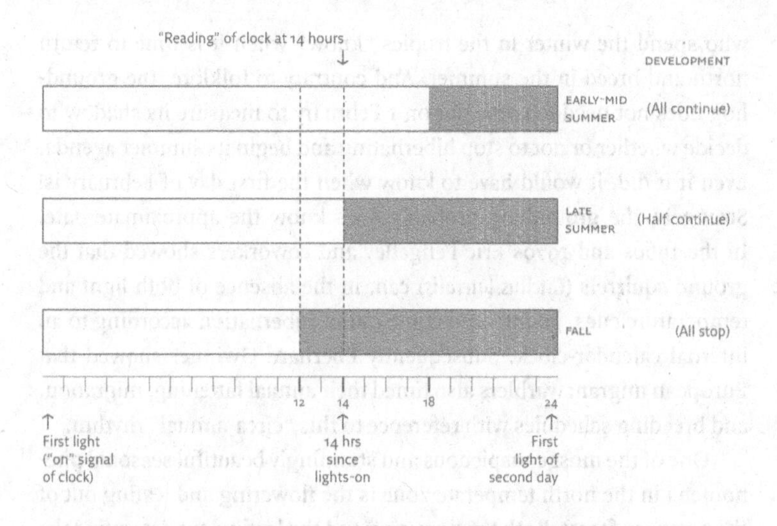


Fig. 3. **How an animal may determine the season by the length of day. Based on experiments with the cabbage butterfly caterpillar, using three different photoperiods.**

when darkness turns to light. They then "sample" for the presence or absence of light after measuring off a specific time period—say, about fourteen hours (the exact time differs in populations adapted to various geographical areas). If, for example, in midsummer the day lasts fourteen hours, then they would "see" light when they sampled at their twelve-hour "window," and then their central nervous system would interpret that as a long day (summer), and continue the normal cascade of hormones to continue their development. However, as the season progresses and the days get shorter, there eventually comes a day when they would experience darkness at the twelve-hour sampling window, and then with no light at that point they would shut down hormone secretion from the brain—until the signal is reversed the next summer, when development would proceed.

Some organisms do not have access to photoperiod signals. For example, at the equator the photoperiod is an even twelve hours of light and twelve hours of dark all year long. Are animals there clueless about what season they find themselves in? Apparently not, because migrant birds who spend the winter in the tropics "know" when it is time to return north and breed in the summer. And contrary to folklore, the groundhog does not need to come out on 1 February to measure its shadow to decide whether or not to stop hibernating and begin its summer agenda. Even if it did, it would have to know when the first day of February is! Strangely, the groundhog probably does know the approximate date. In the 1960s and 1970s Eric Pengelley and coworkers showed that the ground squirrels (*Citellus lateralis*) can, in the absence of both light and temperature cues, go into and come out of hibernation according to an internal calendar-clock. Subsequently Eberhard Gwinner showed that European migrant warblers also timed their annual fattening, migration, and breeding schedules with reference to this "circa-annual" rhythm.

One of the most conspicuous and stunningly beautiful seasonal phenomena in the north temperate zone is the flowering and leafing out of the northern forest. Both the flowering and the leafing out determine the insect populations, which in turn make the summer world possible for the majority of the birds and most mammals.

Flowering and tree leafing are precisely scheduled events. By the end of January we've had three months of seeing all our trees starkly bare, and we're still experiencing snowstorms and bitter cold. "Only four months to go" we think then, before the glorious time arrives when the buds break and the trees flower, and are again resplendent in the long-awaited and much-anticipated color, green!

Our impatient waiting is all the harder when we realize that most of the buds are ready-made all along, just biding their time to burst forth. Indeed, they were already fully formed on the trees the summer before, long before the brilliant leaf shows of early October and the shedding of the leaves a week or two later. Buds are embryonic stems with leaves in one package and embryonic flowers in another (as in alders, hazelnut, and birch), or young stems with leaves and flowers encased all together under the same protective leaflike scales (as in most species). All through winter the various types of buds experience and must survive snow, ice, storms, and thaws, and the tree must bear costs for them to have been produced so early. Grouse live for months almost entirely from eating the buds of trembling aspen and birch. Purple finches, pine grosbeaks, turkeys, and squirrels feed on the buds of maple, aspen, firs, and spruce.

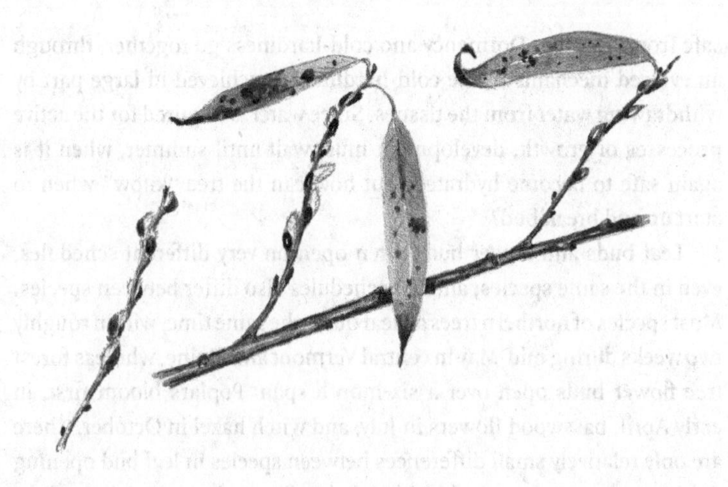

Fig. 4. **Willow twig on 23 October, before the leaves were dropped, showing the "pussy willow" flower buds (along with two tiny leaf buds on the base of each twig, and a portion of the same twig drawn again the following April).**

Red squirrels eat balsam fir and spruce buds (of both leaf and flowers), and indeed they may produce an extra litter of young in apparent anticipation of an episodic spruce cone crop. Although popularly said to be "psychic" and able to "predict the future," they are not the first but are capable of the second—they get their cue from eating the flower buds that precede a seed crop.

Prepackaging the leaves and flowers into buds the summer before they open normally has advantages that outweigh these costs. The main advantage is probably that it helps the tree to flush out quickly, and thereby to maximize the short growing season of about three months. In those three months the trees must not only produce their photosynthetic machinery, the leaves, but also use them long enough to repay their production costs to make an energy profit. Many animals take advantage of the early bud production, but trees are seldom fooled by a false start—which could occur because of a midwinter thaw, making them lose all their investment. As long as the buds maintain dormancy, they remain safe from freezing. Dormancy and cold-hardiness go together, through an evolved mechanism: the cold-hardiness is achieved in large part by withdrawing water from the tissues. Since water is required for the active processes of growth, development must wait until summer, when it is again safe to become hydrated. But how can the tree "know" when to start up and break bud?

Leaf buds and flower buds often open on very different schedules, even in the same species; and the schedules also differ between species. Most species of northern trees all leaf out at the same time, within roughly two weeks during mid-May in central Vermont and Maine, whereas forest tree flower buds open over a six-month span. Poplars bloom first, in early April, basswood flowers in July, and witch hazel in October. There are only relatively small differences between species in leaf bud opening (with quaking aspen and white birch being first; oaks and ash being last; and beech, maples, and many others being in between).

The buds of different tree species each open according to their specific local schedules, which are dictated by a complex interplay of cues involving hours of daylight, seasonal duration of cold exposure, and warmth. Warmth, as such, is not enough. For example, if sugar maples from the north are transplanted to Georgia, they won't break bud there, because they receive insufficient chilling. Their strategy of determining whether or not winter has occurred is like that of the previously mentioned silk moth pupa, which also won't break dormancy unless it (or at least its brain) is chilled for a sufficiently long time.

Although many trees have their primordia for both leaves and flowers packaged into the same bud (for example, apple and other Rosaceae, and viburnums) so that leafing out and flowering occur at about the same time, most of the northern forest trees allocate separate buds for leaves and flowers. This separation of buds appears to be adaptive, because it allows the plant to strategically separate its time of reproduction from the time of leafing out. It thus allows some wind-pollinated trees (the

---

Fig. 5. **Leaf and flower buds of quaking aspen, as they appear from the end of summer to early January, with the flower opening in the first week of February, after being kept warm indoors. Center shows speckled alder twig with leaf, and separate male and female flower buds.**

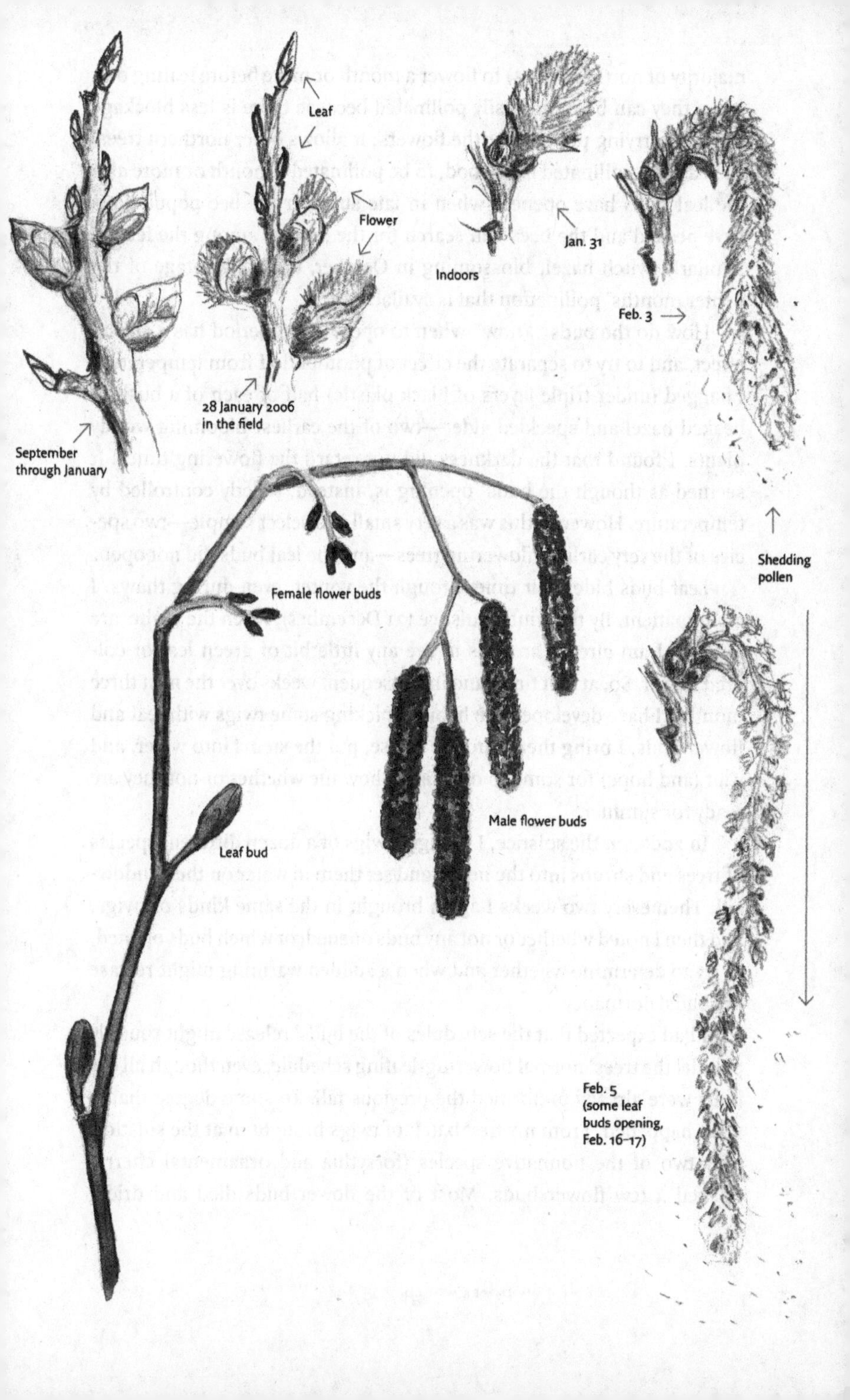
Leaf
Flower
Jan. 31
Indoors
Feb. 3
28 January 2006
in the field
September
through January
Shedding
pollen
Female flower buds
Male flower buds
Leaf bud
Feb. 5
(some leaf
buds opening,
Feb. 16–17)

majority of northern trees) to flower a month or more before leafing out, when they can be more easily pollinated because there is less blockage of wind carrying pollen over the flowers. It allows other northern trees, such as bee-pollinated basswood, to be pollinated a month or more *after* the leaf buds have opened, when in late summer the bee populations have peaked and the bees will search for the flowers among the leaves. Similarly, witch hazel, blossoming in October, takes advantage of the winter months' pollination that is available then.

How do the buds "know" when to open? Photoperiod has a strong effect, and to try to separate the effect of photoperiod from temperature I bagged (under triple layers of black plastic) half of each of a bush of beaked hazel and speckled alder—two of the earliest-blooming woody plants. I found that the darkness did not retard the flowering times. It seemed as though the buds' opening is, instead, strictly controlled by temperature. However, this was a very small and select sample—two species of the very earliest-flowering trees—and the leaf buds did not open.

Leaf buds bide their time through the winter, even during thaws. I am impatient. By the winter solstice (21 December), when the nights are longest, I am already anxious to see any little bit of green leaf or colored flower. So, at that time, and in subsequent weeks over the next three months, I have developed the habit of picking some twigs with leaf and flower buds. I bring them into the house, put the stems into water, and wait (and hope) for some to open and show me whether or not they are ready for summer.

In 2006, on the solstice, I brought twigs of a dozen different species of trees and shrubs into the house and set them in water on the windowsill. Then every two weeks I again brought in the same kinds of twigs, and then I noted whether or not any buds opened, or which buds opened, to try to determine whether and when a sudden warming might release the buds' dormancy.

I had expected that the schedules of the buds' release might roughly parallel the trees' normal flowering-leafing schedule, even though all the buds were already preformed the previous fall. To some degree that is what happened. From my first batch of twigs brought in at the solstice, only two of the nonnative species (forsythia and ornamental cherry) opened a few flower buds. Most of the flower buds died and dried, although the twigs remained alive and some of the leaf buds finally opened in February. But alder, willow, beaked hazel, quaking aspen, red maple, and elm brought in during January opened at least some of their flower buds after only six days. And some of the same species brought into the warmth in mid-March (one to three weeks prior to normal blooming outside) also began to expand or open flower buds in about the same time, three to six days later. As in the field, however, their *leaf* buds remained hesitant to respond to warmth, opening only about a month later. The leaf buds of some tree species, primarily ash, red oak, and sugar maple, showed no response at all even after two months in the warmth.

The restraint in leafing out, although proximally related to potential frost damage, is ultimately probably due to the danger of snow loading that could topple the trees (as discussed later). The risks of frost injury are different for flowers and for leaves. A tree lives for many decades or centuries. It can risk losing its flowers to frost in any one year because energy saved in not fruiting that year can be invested in growth or in fruiting the next year. Losing leaves, on the other hand, results in cutting off energy inflow and stopping growth and hence a falling behind in the competitive growth race for the light.

Tree buds break dormancy owing to local stimulation; prior chilling of one bud on a lilac stem enables it to flower while a neighboring non-chilled bud remains dormant. Similarly, certain chemical vapors applied to one lilac bud will cause it to open while an adjacent untreated bud remains dormant (Denny and Stanton 1928). Therefore, presumably, if a tree is kept in a warm greenhouse all winter it will not leaf out or flower in the spring, although if one branch of it has been protruding to the outside, then that one branch alone will leaf out and bloom. Such simple experiments show that timing—when to renew life for the summer after the long winter—is not left to chance. There are active mechanisms of repression and activation in the development of the buds, and those mechanisms are contingent on cost-benefit ratios. Low temperature plays a large role both in repression and in release, and the timing mechanisms reside in the tissues themselves—not in a central place that then sends signals to the rest of the plant's body.

The twigs bearing buds that I stuck into a jar on my desk, while snowstorms raged outside and Fahrenheit temperatures dipped into the range below zero, and that then produced leaves and flowers, reminded me of summer to come. Beyond that, they reminded me of overeager runners who have prepared for a big race for over six months, and who are ready and set to wait for more and more specific cues that signal the start. The last "go" signal is a warm temperature pulse. Such pulses are sometimes reliable cues of the beginning of spring, but for the leaf buds, apparently only if they occur in late April or early May.

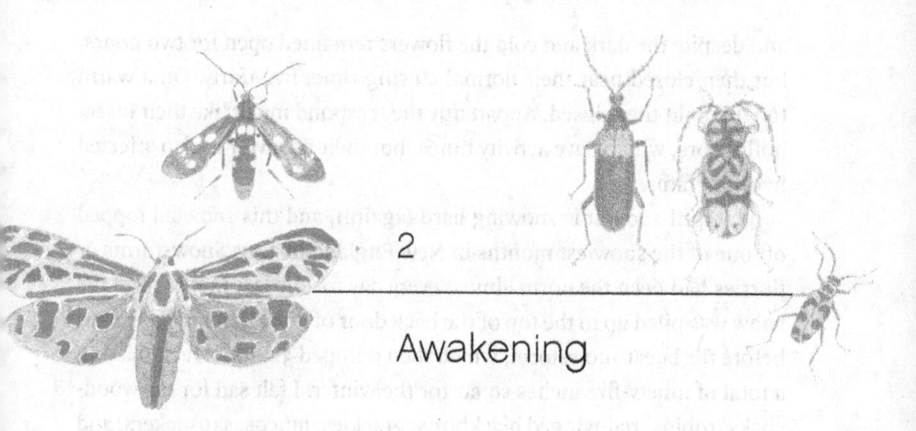

# 2

# Awakening

23 April 2006. THE SUN SHINES AND THE FIRST CROCUSES are blooming around the house. In days, they will be gone for a whole year. I can't resist trying to preserve one by sketching it in color. The flowers normally stay closed all night and open late in the morning as though they awake then. Are they responding to sunlight? Temperature? Time? I observed and experimented and think it may be all of these. The crocuses in sunshine in our yard didn't open until 41°F. If I darkened them (by inverting a garbage can over them) they closed and stayed closed even at 50°F but opened at 70°F while they were still in the dark. However, by five-thirty in the afternoon while there is still sunshine they closed even if at 45°F. Might native flowers act similarly?

I NOTICED THE BLOODROOT FLOWERS FROM OUR WOODS raising their petals straight up at night to tightly enclose their reproductive organs. On the other hand, in the daytime in sunshine the stamens and pistils were fully exposed as the petals were spread out to the sides. However, a day later at 50°F under an overcast sky the flowers were closed all day. I dug up a plant and brought it into the house, and there it stayed open all night, at 60°F. Was flower opening therefore controlled by temperature? I put the plant into our refrigerator at two-thirty PM, and despite the dark and cold the flowers remained open for two hours, but then closed near their normal closing time, five-thirty. On a warm (60°F) night they closed. Apparently they respond much like their insect pollinators, which have activity times, but their behavior is also affected by temperature.

17 April 2007. It is snowing hard (again!), and this snowfall topped off one of the snowiest months in New England history. Snowstorms or flurries had been the norm almost every day for the last month, and the snow was piled up to the top of the back door of our cabin in Maine even before the latest snowstorm, which then dumped 3.5 more feet, making a total of ninety-five inches so far for the winter. I felt sad for the woodcocks, robins, red-winged blackbirds, grackles, juncos, sapsuckers, and flickers that had already returned on their normal schedule, which this year was at the wrong time. Flocks of juncos and robins settled onto the only bare earth available—the shoulders of plowed roads next to tall snowbanks, where there was surely no food for them. How many of these early birds will survive? It might take weeks for the snow to melt. But I was wrong.

Suddenly, "clearing skies" were predicted, and indeed the sun came and along with it a southerly wind. Temperatures soared to 50, 60, 70, and finally 80°F. Rivulets gushed and gurgled down from the hills into flooding rivers. What a difference four days can make: the difference between winter and summer. The goldfinch males that have been coming to our feeder quickly molted their drab greenish winter garb, turning bright lemon yellow in a week. The great greening will soon begin—but not before the wood frogs have had their choruses.

The long-awaited wood frogs were at least two weeks late this year, starting their chorusing on my birthday, 19 April. But the spring peepers were on calendar time, and thus this year they piped up only a day, rather than two weeks, behind the wood frogs. By 23 April, the summer awakening had already progressed far. I was so excited that I could barely sit still to write about it. But I had to do it while I was still reasonably coherent, before the greening onslaught could rush in, and while the impressions were still fresh in my mind.

In only three days after the warming began, crocuses peaked at our doorstep in an orgy of blue, white, and yellow bloom. Growing only along the sides of our dirt road, which was now finally again turning from spring mud to solid summer ground, the coltsfoot suddenly poked their brown flower buds through the soil and opened wide their bright yellow flowers. On the south-facing wooded slopes near our house, spring beauties, bloodroot, and hepaticas were opening their pink, snow-white, and blue and purple flowers to the sun. And the wind-pollinated trees—the quaking aspen, beaked hazel, and speckled alder—suddenly unfurled their tight flower buds to wave them in the warm breezes as though on signal, which indeed is what the warm pulse had been. The elm and red maples blossomed right on schedule as always, although the sugar maple, one of our most common trees and a most beautiful one when it is in full-splendored pale yellow bloom, chose not to flower this year. From Vermont to Maine the sugar maples were barren of flowers (although I found a single tree in flower next to our well in Maine). Willows were slower; they would be two days behind. However, no leaf bud anywhere had so far opened; nor would any open for weeks.

I saw my first orange and yellow bumblebee queen of the season as she was zigzagging close to the ground, as bees do when searching for a nest site. And two overwintering butterflies—the mourning cloak and Compton's tortoiseshell—perched on a sugar maple trunk sucking sugar water at a lick a newly returned sapsucker had made on a tree next to our back door at the edge of the woods. Their wings were outspread to catch the warmth of the sun. "Our" phoebe finally examined potential nest sites, and a tree swallow sailed around the yard, before briefly examining a nest box and then departing. I'm sure it will be back soon with a mate. Early this morning one of a pair of blue jays tore off twigs from a viburnum bush along the driveway and flew off with them into the woods. It has begun building its nest foundation and will soon search for rootlets to line the nest.

Aside from walking around aimlessly and gawking, I have spent the last three mornings comfortably perched on a solid branch of a pine tree growing at the edge of our bog. I tucked myself comfortably up against the thick solid trunk, and leaned back in bliss behind a thin veil of branches that provided both concealment and a view. At dawn, an hour before the sun's glare bleeds the colors, the bog was a study in pastels. There was no green vegetation at all, unless one looks at ground level to spy the blue-green tips of the sedge shoots beginning to pierce the winter-downed brown leaf blades. Aside from the chestnut brown sedge clumps (hummocks) that are surrounded by water, I saw an expanse of beige-yellow cattail swamp with dark brown seed heads that look black in the dawn. The water surface shimmered in colors ranging through black, tan, blue, and dark greenish where the light reflected from the pines at the edge of the beaver pond.

Light reflected from wavelets as muskrats and beavers swam at slow, steady, unvarying speed. Their noses and ears peeked out of the water, etching V's in their wake. One beaver hauled itself out onto an old dam overgrown with viburnum bushes. Its shaggy coat glistened black as it bent over on its haunches and with its front paws brushed the fur on its head and behind its ears. Then it waddled back into the water and slid out of sight. I silently thanked the beavers, because with their dams and their constant cutting of brush and trees they have created this oasis of very varied life in what would otherwise be an almost uniform expanse of forest.

Unexpectedly I hear the resounding "whoosh-whoosh-whoosh" of heavy wing beats, and what T. Gilbert Pearson in 1917 called the "Lord God bird" and we now more commonly refer to as the pileated woodpecker, *Dryocopus pileatus*, lands beside me. In the same instant that I recognize the woodpecker, it recognizes its mistake and flies to the next tree. A pair of these woodpeckers had recently been making their nest cavity in a poplar nearby in the woods. It will take the pair a month to finish excavating their nest hole—one that will next year probably be used by wood ducks or maybe a pair of screech owls or saw-whet owls. Nesting is evident all around. In the distance I now also hear the "baby bird" caws of a crow—and I know them as the sound of a female incubating eggs and begging her mate to come feed her.

A Canada goose gander patrols along the edge of the cattails, and his loud calls echo over the pond. He is responding to another's call, which I hear coming from the distance. His mate remains silent. She is ready to incubate four beige eggs cradled in a nest she has made by pulling the cattail leaves under her while perched on a muskrat lodge. It is her fertile period, and he does not want company in his domestic affairs, especially at this time.

A second pair of geese have started building their nest on the opposite shore of the beaver pond, and this gander ignores them. However, every morning and evening several other geese visit the pond, probing to find an opening. He and the other pair unite to attack visitors, and so far have invariably chased them off. These visitors are highly motivated, as are the defenders. Within several more days it will be too late to start raising a family of goslings this summer. The grackles are much more sociable nesters than the geese. Five pairs of grackles have banded together as a small colony. Every year they nest in the same small section of the cattails close to where the geese nest.

It appears that the geese and grackles know each other as individuals, and I suspect that the red-winged blackbirds are as capable. Both came to the bog in small flocks and will soon nest near each other. Every day they come to our bird feeder in groups of about half a dozen, even after they stake out their nesting niches in the bog. The individual male red-wings have their own little stations or territories, and although they tolerate neighbors there, they gang up on others who come by. The grackles came back to the bog as a group, males and females together. The red-wings also came as a group, but the first vanguard is always all males. The females come weeks later, and they should arrive any day now.

The red-wings and grackles are here constantly, and by now I scarcely watch them anymore. But today I had special visitors: two pairs of wood ducks. I noticed them first as dark formed on the water among the sedge hummocks. They seemed to follow one another, stop, reverse, swirl, and swerve. I don't often use my binoculars, because they greatly restrict my field of vision, but this time I retrieved them from under my jacket. From a distance I had seen no color, but now the females, clad in soft gray plumage, provided a pleasing contrast to the males' bold patterns of red, white, black, purple, tan, green, and blue, a costume so flamboyantly gaudy that it would be hard to dream up. They glistened, and their colors were reflected in the water next to them.

The wood ducks seemed to be animated little robots who swerved erratically into and out of the sedges before aggregating at and swimming around an old abandoned beaver lodge. A mallard drake joined them. His luminescent green head seemed to glow, and he held his head high and turned it this way and that. His soft, barely audible calls sounded like exhalations with a sharp edge. Eventually a female flew in and, while quacking loudly, splashed down beside him. He relaxed then and the two, dipping periodically with their heads underwater and their tails straight up, fed together. After a while another male came near, and the paired male vigorously chased him off. Later on I lost track of the hen, and then I saw the two of them staying together as a couple. I don't know what is going on, but I think the female has a nest somewhere and is laying eggs and he will be mate-guarding her as long as she does, so that the eggs she lays will have his paternity. In a few days there will be no females visible, and then males will again associate with each other.

The swamp is dense and I see only its surface. Much remains hidden, and I hardly know of its existence. Today I had the pleasure of making the rare acquaintance of a bittern. This large bird of the heron family may be here all summer, but one would never know it. Today, however, I heard a bittern's "song," an unearthly sound that carries for miles; one would scarcely attribute such a strange sound to a bird. The bird's colloquial name, "pile driver," is derived from the male's call, which reminds me of somebody driving a stake into the ground with a large sledgehammer in a large echo chamber. I located the bittern's brown streaked form only with difficulty, through my binoculars. Standing among cattails on long yellow-green legs, his erect body with elongated neck and bill straight up, he blended in with the vertical dead cattails. He stood without moving a muscle for perhaps half an hour or more. Eventually he started to creep forward, his every motion epitomizing one's stereotype of the "silent stalker." He hunched over, gradually lifted one leg, and just as slowly in one continuous motion put it down in front of him and lifted the other. Then he stopped, again frozen in position, until, very slowly turning his head, he took another slow-motion step forward, to again stop for a few minutes and then again take a step or two. Suddenly his head shot forward and down with lightning swiftness, and came up with a frog dangling from the bill. If he had not announced himself (to a potential mate) I would never have known he was near. There is much right under my nose that I don't see, and thus I look forward to getting out, again and again—to discovering.

The birds have started their summer schedule. If not in flamboyant garb, then in song, they put a high premium on making themselves conspicuous. Like us, they communicate through the senses of vision and hearing, and so we are fortunate to be able to be spectators. Some, like the male red-winged blackbirds who perch on top of the cattails or bushes, make themselves conspicuous to rivals, and possibly mates, by where they perch, by flashing their brilliant crimson epaulets (which they otherwise can hide), and by backing up their visual display with a vocal one. The bittern can stay hidden and rely on vocal display almost exclusively. But whatever the different animals do, I can't even imagine what summer, and life, might be like without them.

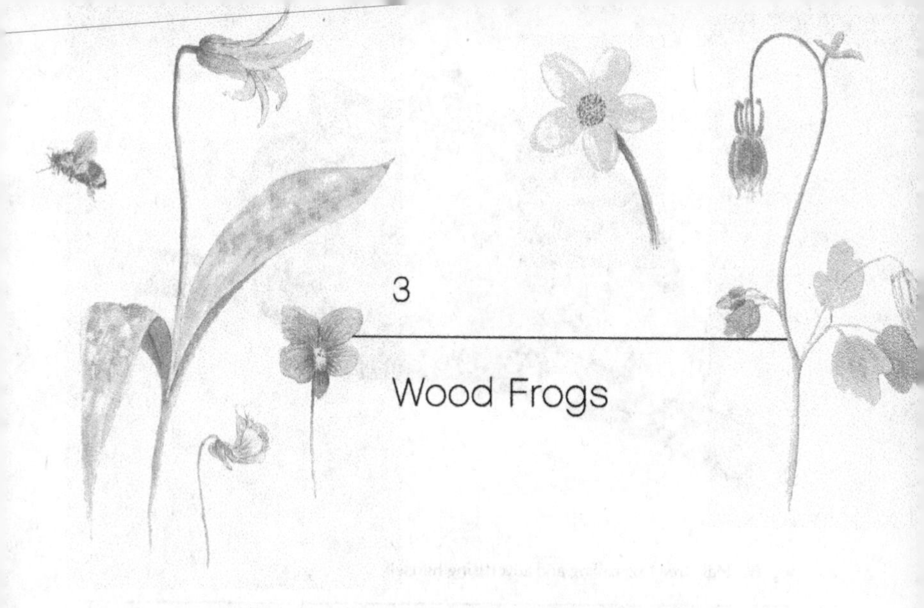

# 3

# Wood Frogs

28 May 2006. IT RAINED FOR A WEEK AND NO INSECTS flew. But today the sun came out and I heard the first gray tree frogs. One male was calling from a branch above the road as I jogged by, and I stopped to find him. He was a gorgeous green (not grayish as suggested by the name). After I climbed up, got him, and brought him home, I put him in a terrarium for a detailed look. He perched on a twig and stayed there like an ornament, but continued to call in three- or four-minute bouts at approximately hourly intervals. When at rest he has a deflated throat that vibrates rapidly at very low amplitude. Then, to call, his whole plump body contracts and suddenly looks skinny, while his loud penetrating churring sound erupts at the same time that his throat balloon inflates.

He produces his loud penetrating calls by exhaling to inflate his throat balloon, and his abdominal contraction is the engine for that burst of sound. With that penetrating sound, his whole body vibrates to its frequency. When my (temporary) pet called, several others within 100 yards of our house joined in. The females, like most other frogs, presumably go toward the loudest, generally nearest, individuals they hear. What a stark contrast to wood frogs that I had watched the month before!

Fig. 6. Male tree frog calling and advertising himself.

IF ANIMALS' MAIN SUMMER PREOCCUPATION IS A RACE of reproduction, then the chorus of wood frogs on a night in early April is the starting gun. The frogs burst out from under the decaying leaves on the ground, overnight meet at a pool that has just melted, and start their convocation, which is rowdy, loud, and brief. One might assume that the males call to attract females specifically to themselves, but now, after getting to know them a little better, I think the story of what they do is more interesting. As we shall see, it can involve cannibalism, and more.

For about eight months the wood frogs crouch, with their heads down and their limbs tucked tightly to their sides, under the leaves that settle on the ground in the fall, and they and the leaves are then covered with snow. The frogs often freeze solid, and in that condition they don't have a heartbeat, breathing, digestion, or activity of the brain cells. A reputable human pathologist, applying the same clinical standards to them as he would to one of us, would conclude that they are dead.

The wood frogs' cue to revive and arise as from the dead, like that of the alder, hazel, and poplar flower buds, usually comes on the first warm (40°F) rainy day in April. By the millions freshly thawed frogs crawl out from under the cool damp leaves, and each of them starts hopping in a beeline to a little pool somewhere in the woods. They arrive at it from all directions. The whole population in any one area will travel mostly at night, and most of the frogs arrive on only one particular night. But adjacent pools are not necessarily on exactly the same schedule. Traffic to a pool on these nights can be intense; up to 4,000 frogs have been counted coming to a single pool in three hours (Bevan 1981).

All fall, winter, and spring the frogs had fasted and waited for their cues to arise and become active. During thaws in January or February, the aboveground temperatures on occasion rose to nearly 60° F, but the frogs still did not budge. Even after they do become active, they still do not feed for some time. First things first. For wood frogs that means sex and egg laying, which they accomplish simultaneously.

I wrote in my diary entry for 14 April 1995 that I had arrived at about ten o'clock the night before at camp in Maine, driving in a drizzle and being impressed by the "traffic." The main traffic on my trip from Vermont that night happened to be crossing the road, and it was mostly

Fig. 7. **A male wood frog in calling position on a pool.**

Fig. 8. **Part of an aggregation of male wood frogs on a pool.**

wood frogs. While coming through New Hampshire I saw them as pebble-like lumps in my headlights against the black, wet tarmac. At one point I was induced to stop my pickup truck, and I caught twenty of them, both males and females. Every one of them had been facing or hopping toward where I could hear a male chorus at full throttle. The road was also littered with the flattened dead—those who had previously attempted to join the gathering throng. From a distance a wood frog congregation sounds like a gaggle of peripatetic ducks. Presumably it is irresistible to female frogs, and I suspect it is to the males as well.

Although the wood frog choruses in different pools are often on their own schedules several days apart, group chorusing is not entirely due to similar arrival times of the participants in any one pool. Getting there on the same night makes it possible to sing together, but that alone does not ensure a chorus. Only the males call, and not at random with respect to one another. Already timed to arrive at the pool within about a day or two, the individuals further synchronize their calling with each other.

The wood frogs' chorusing is, like that of most frogs, an energetically extreme exertion. In their case it's done on a stomach that has been empty since fall. But that exertion is only a prelude for the wrestling matches that commence almost immediately among the males in their attempt to capture females who come to the pool at about the same time. For most of the males that make it to a pool, the first day they get there will be the only one in their lives to mate. Even then, less than 40 percent of them will get that chance. On the other hand, of the females that make it (at outcome about six times less likely for the females than the males), virtually all will have a mate within seconds of hopping into the pool. Each lucky male who does get a female wraps his forearms around her and locks his thumbs together under her neck.

It is almost impossible for one male to pry another off, and the males stay attached to their prospective mates for hours and potentially days (if

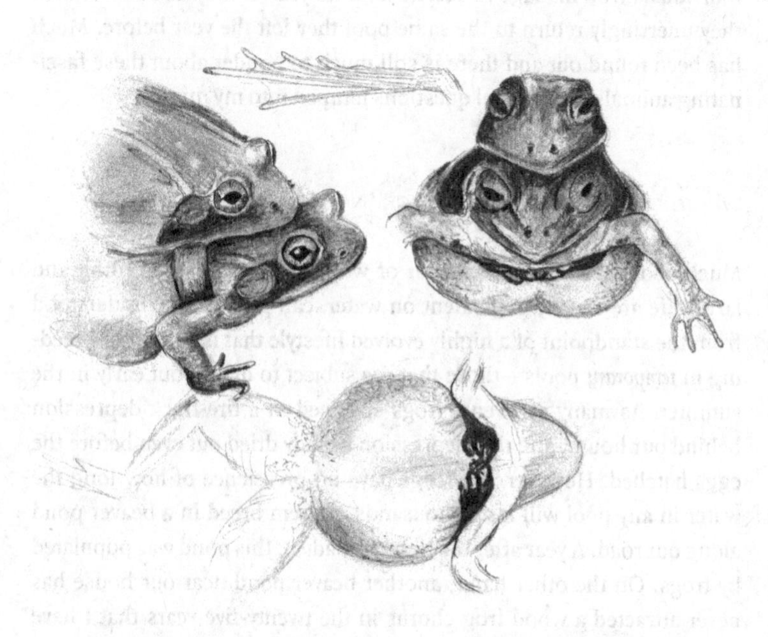

Fig. 9. Male wood frog with neck lock on a female, who will swim with him to the spawning place.

one were to remove them from the pool). A female, even if she is dead, can have a dozen suitors attached to her on various appendages and in various positions. Half a dozen or more males may simultaneously try to lock onto a single live female, but only one of them will achieve the one sure position—a solid neck lock as he perches on her back. He won't let go, and the resulting twosome may look like a two-headed, four-legged mutation; at least one such coupled pair was claimed as such, by a woman who excitedly brought it to my office.

The female then takes the locked-on male for a ride. She swims to the one spot in the pool where all the other frogs will also deposit their eggs. He will not release her until she has laid her walnut-size lump of several hundred eggs, and he then releases his sperm onto them and also releases his grip around her neck. She will then almost immediately leave the pool. This is probably the last time in their lives that either will ever have contact with open water again, except for those rare and lucky individuals who manage to survive another year. If they do survive, then they unerringly return to the same pool they left the year before. Much has been found out and there is still much to ponder about these fascinating animals, but several questions jumped into my mind.

## Where and Why Do Wood Frogs "Nest" Communally?

Much about the unique behavior of wood frogs, whose breeding and larval life are strictly dependent on water, can probably be understood from the standpoint of a highly evolved lifestyle that is suited for breeding in *temporary* pools—those that are subject to drying out early in the summer. As many as twenty frogs spawned in a tire-track depression behind our house, and this depression usually dried out even before the eggs hatched. However, the frogs have no prescience of how long the water in any pool will last. Thousands of them breed in a beaver pond along our road. A year after the beavers made it, this pond was populated by frogs. On the other hand, another beaver pond near our house has never attracted a wood frog chorus in the twenty-five years that I have been watching and listening. It is, however, an ancient (i.e., "permanent") pond, one that is populated not only by all the other local frog species, but—more significantly for the wood frogs—also by minnows, sunfish, and catfish. Wood frogs even breed every year in a washtub-size pool formed in a rock depression on a hill at my camp in Maine. This particular pool is not subject to dropping groundwater levels, and I've never seen it dry out, but fish have never reached it. In short, wood frogs avoid breeding in water that contains fish; and the smaller the pool, the more likely it is to dry out at some time during the summer and therefore to be fishless. Pool size, permanence, and impermanence, as such, make no difference. Apparently, adult wood frogs have an aversion to fish, and for good reason. Wood frog tadpoles have the bad habit (relative to other frog tadpoles) of swimming around conspicuously near the water surface and feeding there on algae, rather than hiding on the bottom like most other tadpoles. I released a handful of wood frog tadpoles into an aquarium containing native fish that immediately went into a feeding frenzy and ate every single one.

It may seem odd that although the frogs don't care much about size, and perhaps other physical attributes of a pool or pond in which they spawn, they nevertheless are very particular about *where* in the pool they spawn. The females make an effort to deposit their eggs at the same place where others have already done so. The males are also nearby, gathered for their chorus. But you need a large pond to notice this. Three miles down the road from our house is a pond about 660 feet long and 165 feet wide. It has ample space for frogs to spread out, yet every spring the wood frog chorus is restricted to an area of several square yards on one end, and almost all the many females drop their eggs there, in one big pile. Why?

Although an individual egg clump is only walnut-size when deposited, it swells to the size of a baseball or softball within hours because the gelatin surrounding each egg absorbs water. When hundreds of individual egg clumps are deposited next to each other, there is a solid expanse of jelly densely dotted with the black eggs (they are white on the bottom, and they right themselves if turned). For female frogs it presumably feels right to swim to where others have spawned or will spawn. We might say that they respond to a "social releaser." And for a male, the releaser to release his sperm is probably the female's releases of her eggs. The frogs do not know the connection between their acts and the ultimate or evolutionary sense or consequences of these acts. Nor do they need to.

It seemed to me that the wood frogs' egg aggregation might have something to do with elevating egg temperature to speed up hatching rate. Using my electronic thermometer, which I had then been using mainly to measure the body temperature of bees, I immediately got busy, waded into the icy waters of many pools, and measured the temperatures around the edges. I found no evidence that egg clumps are located in warmer parts of the pools relative to other parts, so the frogs did not search for or find any hot spots in their pools. But temperature could still be important in egg placement.

The upper black surfaces of wood frog eggs must absorb heat in sunlight, but the heat would normally be quickly dissipated by convection to the cooler water around them. However, a large mass of jelly embedding the eggs would reduce water movement and could aid heat retention. The larger the effective mass, the less heat loss from the center. So, to find out if there is a measurable effect, I took dozens of egg clump temperatures versus surrounding water temperature, and compared single and clumped egg clusters. The results: Egg clusters in shade are nearly the same temperature as that of the water surrounding them. However, in sunshine, the single egg masses were heated on average 3.5°F above the surrounding water temperature, and clusters of ten or more egg masses were heated 9 to 13°F above water temperature.

Temperature affects the developmental rate of the eggs. To find out how much, I brought egg clumps into the lab to determine the time until they hatched. In the woodland pools I had measured egg mass temperatures of 43°F to 79°F, and in the lab none of the eggs hatched that were held either below 41°F or above 86°F. In between these extremes, the developmental rate of embryos was directly related to temperature. For example, eggs at 46°F required thirteen days to hatch, and eggs at 68°F hatched in six days. Thus, within the physiologically suitable temperature range, every increase of 3°F in egg temperature speeds up the hatching date by a day. That could be a huge potential advantage, since the snowmelt pools that these frogs use are highly ephemeral in the summer.

After finishing my temperature measurements I was both pleasantly surprised and also a little disappointed to learn that I had indeed been on the right track. Egg clumping for heating was an idea that had already been proposed. Bruce Waldman from Cornell University had published a detailed study a decade earlier, where he found that the edges of wood frog egg masses were heated several degrees above water temperature while the centers warmed to over five F higher.

### Why Do the Wood Frogs Call?

If there is one tenet that I knew beforehand to be firmly established in the scientific literature on frogs and toads, it is that males compete in attracting mates by making conspicuous vocal displays, and that females choose. Like the singing by birds and many insects, most famously crickets, katydids, and cicadas, the calling by frogs is an advertisement in which a male draws attention to himself or to some resources he holds that females need for reproduction. I know of no published exception to this explanation for male frogs' mating calls. Biologists agree that calling sets up competition among males, allowing females to sort them out and choose. If several calling males are near each other, they can presumably be all the more easily compared, so that females can exercise even better choice. Indeed, aggregations of males where females come to be mated are considered the equivalent of mate marts where the males strut their stuff (if they can, or if they have enough to be allowed by the competition, or both). The females generally choose a small number of individuals out of the participants. Do wood frogs? My hunch was that they do not. And this time I consulted the literature as well as the frogs.

Three research papers on wood frogs' mating aggregations appeared between 1980 and 1985. The first one, by Richard D. Howard (1980), then at the University of Michigan, established that the males outnumber the females at the breeding pools by about six to one. The skewed sex ratio apparently results from different mortality; fewer females live long enough to reproduce because it takes them one year longer than the males to become sexually mature. Each female was found to pair with only one male, and vice versa—a condition that is, I think ironically, described as "monogamy." There is an intense competitive scramble among the males, but Howard was unable to demonstrate any choice by females. A nearly simultaneous study by Keith A. Bervan (1981) also reported no evidence for female choice. Bervan also found that any female would be clasped long before she reached any calling male that she might choose.

The males don't have the luxury to choose, either. Bervan noted that they attempt to clasp with each other, with any females, and even with already firmly clasped pairs. That is, males cast a wide net, try to capture a mate first, and discriminate later; those males on the receiving end of a "trial embrace" give a call that identifies their gender, and they are then immediately released. However, they are not so easily deterred from females who already have a male attached to them. If the female doesn't make a quick enough getaway after she has a male, she will quickly accumulate a surplus of males that restrict her mobility. The other males try to grasp her around the abdomen and then move upward in an attempt to pry the other male off her back. They rarely succeed, but they can do so if the competing male is small enough and the female is so large that her male can't reach all the way around her neck to secure a solid lock with his thumbs.

If these two studies weren't enough to dispel the notion or expectation of female choice in wood frogs, another one did. This next study, by Richard D. Howard and Arnold G. Kluge of the University of Michigan (1985), emphatically concurs with the previous studies. These authors write, confidently: "Our results were unambiguous: the slightest movement by females resulted in their immediate amplexus [locking on of male] by the nearest male." And females did not dislodge those potentially unwanted males; only other males did that, in athletic wrestling contests where size and strength mattered. Thus the most exhaustive and broadest study of "mate choice" yet (involving reams of data on survivorship, growth rates, numerical estimates of zygotes produced by females and sired by males, and monitoring of 5,877 individually identified frogs) still had found no evidence of mate choice in wood frogs. My hunch, derived from one glance into a pool, had apparently been right, but hunches as such seldom win kudos.

It may seem that there was little left to learn about the mating habits of wood frogs, given the solid empirical results. However, I still wondered: If females don't choose, then why do males call at all? What could they call for?

To answer this, or any other relevant biological question, it helps to look first at the context in the field, in the animals' natural environment. Wood frogs are unlike any of the other species of local frogs in that wood frog males are not spread out. By contrast, male tree frogs are separated from each other by trees; spring peepers, green frogs, bullfrogs, and other pond breeders are usually scattered along a shoreline or over an expanse of marsh where they can be hidden in little niches under leaves and grass that allow them to control space around themselves. Calling wood frog males are easily visible, massed on open water near the center of their little pool.

Several days after I had seen the wood frogs crossing the road at night while I was driving to Maine, I sat down near the edge of a pool near my camp. This pool, which is no larger than the floor space of an average room, contained at least fifty highly visible male frogs. They were spread out, as is typical, about a foot apart all over the surface of the pool, with only the tops of their heads out of the water while their hind legs trailed behind. They floated in place and occasionally paddled with alternate strokes of their hind legs. They approached any other frog they came near. I saw only one female jump into the pool. At least I assumed it was a female, because only this one was pounced on almost instantly and not released. In seconds, three males were on top of her, and one of them got a tight neck lock. It was, as always in wood frogs, a classic competitive scramble with the males in an intense contest for the females, who are literally up for grabs.

In one ball of ten squirming males that I untangled I found a dead female at the center. I threw her back into the pool, and she was again mobbed and embraced in the same way. I suspect that there were so many males on this dead female because she could not escape. But males' preference could also be involved, because males "should" prefer more rotund females: such females would make the males' sometimes prolonged wrestling efforts more worthwhile, since they would get more eggs with one shot (so to speak). My dead female happened to be rotund indeed—she was bloated with gas, though, instead of eggs. In any case, she could not have chosen any of these males; they all chose her. Whatever the calling behavior of one or the other of these amorous males, it had made no difference to this particular long-dead female. However, an anthropomorphism readily suggests itself to describe what might be going on. Is the frogs' chorus a collective effort of the males to get females to come into their pool, like guys at a Saturday night college fraternity party playing loud music to attract the most babes to go to their house as opposed to a neighboring one?

At first I watched the frogs from a distance of about twenty feet, so as not to disturb their activities. Their calling was, as usual, in concert; some of the time the whole crowd seemed to be sounding off, and then there were periods of silence, as though the band played all together for most effect, then took an occasional break before resuming with renewed vigor. After a while, one or two of the frogs started up again, the rest then joined in, and their voices blended in. (By contrast, with many other frogs and toads one can easily pick out individuals by differences in pitch.) I had brought along a tape player to record the chorus. I reversed the tape and played back their calls during a silent period after I had disturbed them and they had dived to the bottom. Almost instantly after I turned on the sound, frogs started popping up to the surface and chimed in with the taped sound. I then shut the sound off, and then they stopped too. When I again played the tape I got the same result. I repeated the trial fifteen times, and it always worked.

Like most summer activities, the frogs' vocal signaling requires an impressive expenditure of energy (Taigen and Wells 1985) and therefore presumably has an advantage. However, it is not immediately obvious why the calling of an individual wood frog in a chorus could aid him snag a female that has jumped into the pool. If not, then why call at all, as long as other nearby males are doing all the work and bringing females in by their calling? Instead of having a mating advantage, calling would seem to be disadvantageous, because the noncallers, who save their energy, should have an advantage in the inevitable tussling contests that ensue among the males seconds after a female jumps into their crowd. In the extensive literature on the mating game there are indeed innumerable examples of "satellite" males (those that wait to intercept females coming to the displaying males they are attracted to), who adopt the energetically more economical mating strategy. So, why don't they all stay silent? And after one frog called, then all the satellites should especially be silent. Instead, all the neighbors joined in. It didn't seem to make sense from the perspective of satellite males. But I knew it makes sense—somehow.

Although the frogs' synchronous chorusing in crowds was puzzling to me, someone thoroughly steeped in frog literature would not have been confused. One reviewer commented about an article I had submitted that was summarily rejected: "Why of course they would chime in. Frogs do that. When your competitor is calling you had better immediately call as well to remain in the picture as far as a potential mate is concerned." This idea makes intuitive sense, of course, but only in terms of females' choice. I had regularly observed exactly that—in other frogs, those that are hidden and widely spaced, such as gray tree frogs, spring peepers, green frogs, and maybe bullfrogs (although those in our pond also aggregated to produce deafening pulses of sound where hundreds joined in, separated by moments of absolute silence). Why should male wood frogs chime in with the chorus of voices when there is no way for an attentive female to choose specific preferred individuals out of the chorus line? I didn't know. But I didn't think they were performing like the participants in a summer music festival. Or were they?

In a scenario of intense male-male scramble competition for females, the idea that these frogs could be "cooperating" is not intuitive. But by calling synchronously they may be. This is because the communal pool din, which is what the females respond to, would offer a louder signal and would thus attract more females. The males would, of course, not eliminate competition among themselves, but they would reduce it since the ratio of females to males would increase at their pool relative to a neighboring pool where the males do not amplify their attractant.

## How to Describe the Young Frogs' Summer Race?

Regardless of ultimate cooperation by males in attracting mates, northern wood frogs breeding in woodland puddles live on the edge of survival and proximally compete for their lives. In their ephemeral pools, wood frogs have only about two months of summer to complete their larval development. Often they run out of time. In 1995, twenty-one of the twenty-four pools I was watching were dry by the beginning of July. (The next two years brought the wettest springs on record in Vermont, and none of the pools dried out.) In 1999 all my pools were dry by May 18; it was one of the driest springs on record. But as I will show, cannibalism then saved some froglet cohorts from annihilation.

After getting the earliest possible start following the spring melt, the next step in the wood frog's race with time is to get the tadpoles developed into frogs and hence out of the pool before it dries up. This step mainly involves the development of the larvae; they must grow fast, or be able to act like adults (hop on the ground; breathe air), or both. Two main ingredients make the difference in this race: food availability and temperature. Any increase in body temperature above the near-freezing water temperature that the tadpoles find themselves in permits an increase in growth rate, and they must simultaneously have access to sufficient amounts of the right kinds of food, particularly protein.

Wood frog tadpoles are primarily vegetarian. They feed on algae. But algae are not their only food. When I kept tadpoles indoors in an aquarium, adding nothing but a few decaying leaves picked out of woodland pools, they fed on these leaves and skeletonized them so that only a fine latticework of leaf veins remained. Larvae fed on dead leaves were still alive at the end of February, when I found them under the ice of a rain barrel where I had dumped eggs the previous April. Apparently an insufficient (low-protein) diet can extend the tadpoles' life span from several weeks to at least ten months. Conversely, when I fed them some "fish flakes," a high-protein commercial aquarium diet, they had a growth spurt.

Freshly metamorphosed froglets look and act like adults. However, they weigh as little as 0.007 ounce (0.2 gram), about one-hundredth of the adults' mass. The specific size of the tadpole where evolution has set the developmental switch to produce the adult form is flexible, but in wood frogs it has presumably been strongly influenced by time. The smaller the tadpole at the set point, the shorter the time to get there. Still, in many years the larvae run out of time and don't make it to the froglet stage. When I checked on one of my biggest study pools on 11 June 1999, I found the center a moist black goo of dead and dying wood frog tadpoles. It was surrounded by the tracks of raccoons and great blue herons. Carrion beetles and maggots were mopping up the edges. Similar scenes were repeated at other wood frog breeding pools, but as I found out later, this did not mean the wood frogs' breeding at those pools was necessarily a failure that year.

I scooped up a few spoonfuls of the dead and dying wood frog tadpoles and dumped them into my aquarium with their living relatives. The larvae immediately consumed the dead and weakened of their own kind, and literally overnight they grew hind legs. The next day they added front legs as well, and the following day their tails shrank. They still swam like tadpoles when in the water, but they hopped like frogs the minute they were on land. In three days the change to a cannibalistic diet had caused them to become frogs, whereas the other larvae of the same batch who were maintained on decaying leaves were still tadpoles seven or eight months later. I suspect therefore that as the pools dry up in the summer, feeding frenzies on their own kind impel some froglets to life on land.

Initially a pool where no more than 100 females have deposited their clutches of 300 to 1,000 eggs each may contain at least 50,000 tadpoles. Collectively they represent the nutrients that have been concentrated from highly dispersed and often microscopic food particles by the collective grazing of the tadpoles over the span of a month or two. They become a nutrient store that could later give their brethren a boost to lift them out of the pool as the water disappears. If the same scenario is repeated regularly through evolutionary time, then cannibalism could be an important part of this frog's survival "strategy" (a response resulting from evolutionary selection).

Temporary pools are a prime component of the wood frogs' summer world, and I conclude that almost everything they do is highly evolved to take advantage of it. Their specific behavioral mechanisms blur the meaning of, or give new meaning to, our ideas of "cooperation" and "competition."

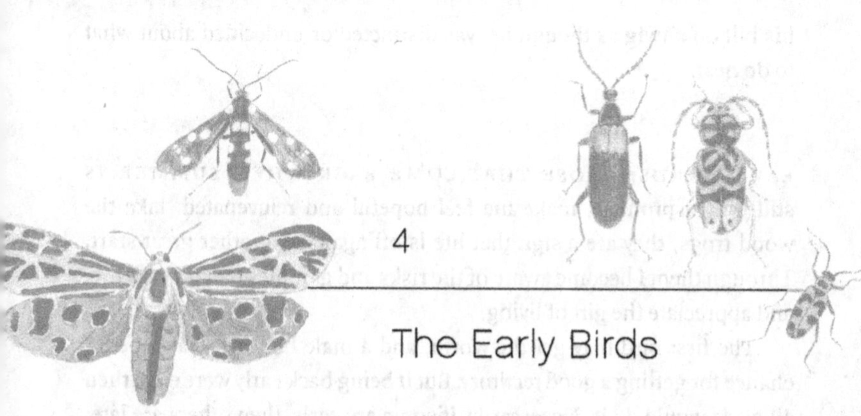

# 4

# The Early Birds

11 March 2006. IT'S ALMOST TIME FOR THE FIRST BIRDS to start coming back, and I've gone into a frenzy with handsaw, hammer, and nails, making nine birdhouses. I hung them around our house, thinking of wrens, tree swallows, and maybe bluebirds. My timing is right. By eight AM the first red-winged blackbirds arrive below in the beaver bog. I see four of them "on station" calling "oog-la-ee" from the tops of bushes and cattails. After an hour they fly up the hill to our house and land on our bird feeder. Last summer they fed here on sunflower seeds, which are not even visible from the outside and can be reached only through a little crack. These birds today act as though they are familiar with the feeder situation, and I suspect that some of them are the same bunch every summer. In the next several weeks they will come to the feeder every day, and they often come in small groups; they are still flock birds even though when they fly back down in the bog they spread themselves out. Down there they remain within visual and vocal contact of each other. Grackles also came back today—I saw three circle the bog. Later a relatively tame one came up to our house and perched on a big black cherry tree that shades the front of it. In hopes that it might be Crackle, whom we had rescued last summer from a nest that was crawling with mites and who became our friend and the favorite pet of the kids, I called "Crackle! Crackle!" He didn't fly off. Instead, he wiped

his bill on a twig as though he was distracted or undecided about what to do next.

EARLY BIRDS, THOSE THAT COME BACK WHEN SUMMER IS still only a promise, make me feel hopeful and rejuvenated. Like the wood frogs, they are a sign that life is off again to another great start. Through them I become aware of the risks and gambles that go with life, and appreciate the gift of living.

The first bird may get the worm, and a male bird also has a better chance for getting a good territory. But if being back early were easy, then all birds would do it. Necessarily, if some are early, then others are late, in the same way that there is no winning without losing. The benefits of being early have to be balanced by the costs, or all birds would be early. And there are great costs—the possibility of foul weather and lack of food, both of which kill.

As I'm starting to write this, again in mid-March but a year later (2007), it is time once more for the first migrants to return, but temperatures are dropping and the meteorologists predict twelve to eighteen inches of new snow over the next few days.

The forecast was correct. Then another snowstorm came shortly on the heels of that one. This year, many early birds would have starved. Flocks of dozens to hundreds of returning juncos were by the roadsides along the snowbanks in the Maine woods near my camp. I sank up to my thighs in the snow in the woods next to them where they would normally have been replenishing their spent fuel reserves. Later, in the summer, I saw none of these birds near where they are commonly summer residents.

The average timing of the different species necessarily differs. Bud-feeders and eaters of seeds and berries that stay on the trees can stay north all winter; but those, like the juncos, that feed on the bare ground must leave in the fall after the first snowfall and can return only after the snow melts. Those that feed on flying insects come next, and the caterpillar-hunters can't risk coming back until after the trees have put on their leaves in late May or early June.

*The Red-Winged Blackbirds.* Flocks of red-winged blackbirds normally make their first probes into their northern breeding grounds during the first week of March. Twenty to thirty males perch close together high in trees along snowed-in bogs. They are then already fully attired in their ink-black nuptial plumage, with a fluorescent-crimson epaulet ready to display on each shoulder. However, as long as they are in the flock they almost completely hide their flashy badges with other feathers.

The flock, when it first comes to our marsh, ceaselessly chatters high in the top of a maple or some other tree that is still bare. Eventually one or two of the more eager or venturesome birds will fly down to the willows. Another one or two will follow. Then, for the first time, you hear their unmistakable vocal signature, the "oog-le-eee" that they give only at "home." Then also, for the first time they display their garish crimson shoulder patches, which they have so far kept hidden. The males must be displaying for each other; the females won't be back for weeks.

Within half an hour the whole flock may reassemble in a tree, and then fly off again. But from now on they will reappear almost every day, and each time these males will spread out in the bog and take up their stations at specific cattail stalks or viburnum bushes. Each day they come a little earlier from feeding areas in surrounding fields and woods where the snow has melted. By early April, when the pond ice melts, they begin to stay almost full-time, and by then (if not since years before), they probably know each other. Latecomers, who are probably strangers to the bog, are chased vigorously, not only by any one territory-holder, but also with the active participation of his neighbors.

Then one day the sun shines brightly. The ice melts. The first painted turtles come out of the mud and sun themselves on half-submerged logs, a bittern calls from his hiding place among a tangle of last years' cattail fronds, and a snipe who seems only a speck high in the sky sounds forth in an unearthly whinnying. Now the redwing females, brown sparrow-colored birds, arrive and skulk close to the ground amid the tangled sedges and cattails. Then, after the sedge leaves start poking straight up through last year's matted brown leaves, you may—if you are patient—see one of the unobtrusive females carrying in her beak a long-dead brown sedge frond. A nest is being built; ovaries are enlarging and eggs are ripening. One after another of four eggs, with a sky blue shell marked with purple and black squiggles and spots, will be sheathed within the oviduct. The female will sit in the nest for four mornings in a row, to lay for four days, one egg at a time. During those days, or just before, you see her making a "baby bird" display by vibrating her wings, and then you see a copulation.

*The Woodcock.* The woodcocks arrive on the first patch of earth that's clear of snow, in late March or early April. This is also when the geese first return for a visit to see if the pond is free of ice. It usually isn't, and they walk around on the ice and then leave, to try again a few days later.

The woodcocks (also called timberdoodles or wood snipes) put on spectacular flight displays that commence almost immediately after they return. A woodcock obtains its diet of angleworms by probing in the mud with its specially designed tool, a long bill with an overhang of the upper mandible at the tip. Do woodcocks probe at random? How or even if they feed when they first come back (when the ground often freezes solid nightly) is a mystery. Aside from food, what a male certainly needs on first returning is a lot of sky for his mating dance, and a little patch of open ground as a landing and launching platform.

Fig. 10. Portrait of a baby woodcock, showing its already well-developed bill, which is specialized to probe deeply in soft mud.

Words cannot do justice to the woodcocks' sky dance. As a prelude to it, the woodcock, with a puffed-out chest that makes him look like a miniature bantam rooster, struts on his little patch of overgrown field and makes little hiccup sounds interspersed with "peents." He gives the impression of a drunk on parade, but then he takes off like a rocket with whirring, whistling wings. Off he flies in a straight line upward, and then after gaining altitude over the treetops surrounding the clearing where he started, he begins to ascend, in ever tighter spirals, into the sky. You hear a high-pitched whistle—made possible by three stiffened feathers on each wing—pulsing to his steady wing beats, sixteen times per second. Then, after reaching an altitude where, although he is the size of a robin, he looks like a tiny dark speck, he interrupts his wing beats with rhythmic pauses and fills in this momentary silence with a high-pitched rhythmic vocal tweeting. The tweeting becomes louder and louder, and the wing beats are more pulsed and rapid, until a crescendo is reached, and at that point he begins his final approach earthward, diving from the height that had made him appear to be but a speck against the darkening sky. His wings still beat rapidly, but his sound-producing feathers are somehow decommissioned so that all you hear now is a dull flutter just before he lands. He settles down at almost at the same spot where he started, to once more resume strutting, hiccupping, and peenting.

The woodcock's sky dance dazzles because it is both spectacular and subtle. I cannot imagine a summer beginning without it. The sky dance evokes memories of fishing trips to Enchanted Pond with my friend and mentor in Maine, Phil Potter. We camped along the shore opposite a great golden eagle nest on a cliff rising from the opposite shore, shortly after the ice went out. We sat next to a blazing campfire under the stars, and heard the birds' sweet refrain in the background from somewhere far away. I've never tired of experiencing such raw enthusiasm that seems to knows no fatigue, no diminution. I've lain under the moon to hear the performance when I went to sleep and again as I woke up. Sometimes I'd find my sleeping bag covered with fresh snow, and I'd wonder if any of the woodcock hens might already be sitting on four yellowish tan eggs that are spotted and mottled with brown and magenta and blend in, like the hen's back feathers, with last year's dried leaves.

*The Phoebe.* A late March snowstorm earlier in the week dumped inches of snow on us, but a south wind is melting it fast. A robin sings, and the redwings are yodeling down in the bog. I expect the phoebe back at any time, too. The phoebe would be flying north now, aided by a wind at night from Alabama or Georgia, and powering itself to hurry along on the homeward journey back to a mere pinpoint on the continent—the house where I live and from where it had left to go south last September. Such feats of endurance and navigation are routine for many migrant birds, but how they might be accomplished still boggles my imagination, no matter how many "explanations"—such as magnetic orientation, use of landmarks, solar orientation, precise timing, and use of prevailing winds—are or could be involved.

I wake up in the gray dawn to the sounds I've long awaited: a loud, emphatic, endlessly repeated "dchir*zeep*, dchir*zeep*." The bird's enthusiasm is infectious. I jump out of bed and announce, "The phoebes are back!"

"The" phoebes leaves a lot unsaid. I have been intimate with phoebes since 1951, when I first met a pair on our farm in Maine, and in our outhouse admired their mud nest, which was garnished with green moss and contained several pearly white eggs. Although I once saw a phoebe nest on a cliff in Vermont, phoebes now nest almost exclusively on and in human dwellings. In the northeast, almost every homestead in or next to woods hosts a resident pair. Phoebes are a fixture of nearly every old farmhouse, barn, or sugar shack.

After I jumped out of bed I took a good look at our friend. There he (I assumed) was, perched on a branch of the sugar maple tree, about six feet from our bedroom window. He was dipping his tail up and down, a phoebe gesture signaling health and vigor. As I watched this sparrow-size bird from up close, I noted his black cap, white throat bib, and dark gray back. He stretched a wing and shook his fluffy plumage, and I felt transported, as if into another being. I experienced a glow of warmth and satisfaction, as anyone would when confronting a marvel of creation that magically appears on one's doorstep at almost precisely the time that one predicted it would come.

Already in the gathering dawn the phoebe is inspecting the two potential nest sites on the house: a one-inch shelf under the roof near the back door, and the bend of a drainpipe near an upstairs window. Now, as he inspected each of these sites, he was making soft churring calls and excitedly fluttering his wings.

On the next dawn he called continuously in the typical phoebe song—a short "fee-bee" alternating with "fee-bay," at his typical tempo of about thirty phrases per minute, repeated with clocklike regularity. He

Fig. 11. **Phoebe at its nest on a board I put up inside our chicken shed. The speckled egg is one that a cowbird had dumped in.**

called from the very top of a big maple tree, then flew over the forest in the direction of our neighbors' house. I assumed this was a male recruiting a mate. Indeed, before the end of the day there were two birds around the house, and two days later they chased off a third. Both were then still inspecting the two potential nest sites.

By the third morning the pair were paying particular attention to just one nest site. They had chosen the thin shelf under the roof by our back door, which we use as our main entrance.

Further nesting progress was then suddenly interrupted. For a whole week there were dark skies and a drizzly rain that turned to snow. Both birds became silent, and then after a couple of days they became lethargic and fluffed themselves out. Soon their wings drooped rather than being folded tightly over the back as they had been before. There were no more flies to be had, at least not by the phoebe's usual mode of hunting, which is to sally forth from a favorite perch to snag those buzzing by. There was no chance that any insect would fly by in a snowstorm. I wondered if the phoebes would survive. To my great surprise, one of the birds hopped like a sparrow onto the snow-free ground under my parked pickup, perhaps to try something different. It also hovered in front of the suet I had set out for woodpeckers, nuthatches, and chickadees, and eventually fed from it. How did it know this was food? There could be no genetic programming in phoebes to feed on suet. Maybe they had taken their cue from the other birds they saw feeding there—a possibility that could account for the behavior of many different kinds of birds with vastly different foraging techniques who all exploit the exotic food and feeders we provide for them.

When the weather improved, the phoebe pair again looked and sounded cheery. As before, they churred and at the same time fluttered their wings in apparent shivers of excitement when they perched at their chosen nest site. The phoebe's song would from now on be repeated thousands of times like a mantra, and it brightens my day even before my morning coffee. I don't know why a song that is as monotonous, unmusical, and undemonstrative as this one still has such a cheering effect on me. It can't be the virtuosity of the performance. The phoebe is one of the passerines, or perching birds, the most successful (that is, the most diverse and numerous) on the planet. The passerines are divided into the songbirds—technically, oscines—and their less musical brethren, the sub-oscines, which include the phoebe. It's a seemingly plain bird in all respects.

Although phoebes are not known as vocal virtuosos, they make many different sounds and gestures that are related to context and that evoke emotion. When "on territory" they begin calling vigorously before sunup, and they then become almost silent half an hour later. When mates find a nest site and show their enthusiasm with soft comfort calls, they come to a consensus or agreement with each other. When the adults start to build the nest, they also "chip" to each other, occasionally throwing in an excited "zeebit" or a "chirreep." Similarly, the laying of the first egg and the hatching of the first young both induce excited vocalization. Several times I have heard excited "singing" at midday, and realizing that there was something unusual, I checked and found that the young were starting to hatch. Coincidence? Possibly. I have also heard similar excited singing after I removed a chipmunk that they were mobbing, which had been trying to get at the nest. There may be a common connection between all these very different events that induced the same emotions.

The young's begging "cheeps" in the nest are barely audible to us (presumably the low volume reduces the chance that they will become an advertisement to predators, since birds in safe nest sites, such as woodpecker young in solid tree holes, are almost always continuously noisy). After the young leave the nest the parents "chip" and the young "cheep" in answer, but they call more loudly now so that they can be found and fed. Their varied vocalizations are not words. They convey emotions that I may not feel as they do, but that I can understand.

In the first week of April the presumed female (males and females have identical garb) began carrying mud in her bill from a puddle in the driveway. She plastered it onto the thin ledge, the edge of a board, under our back porch. She also brought back green moss from the woods, and reinforced, decorated, and camouflaged the nest with it. Later, after the nest cup was finished, she picked up stray dog hair and grass fibers to line it, and then in the last week of April she laid one immaculate white egg a day, until she completed a clutch of five. By this time she had become used to our comings and goings and would rarely flush from the nest. The eggs hatched after a two-week incubation. The yellowish pink chicks were covered with only a few sparse plumes of white fluffy down.

By early June, when the young were almost ready to leave the nest, I saw one parent come with a big grasshopper in its bill. Only one baby gaped; the brood must have been well fed. Then there was a downpour, and immediately afterward I heard an animated "phee-bee" song. But this one was not delivered from a perch near the nest, as usual, nor was it coming at dawn, the usual time for song. It was, instead, coming near dusk. I looked up and there he was like a skylark or a woodcock circling high in the sky, but only for a few moments. Almost immediately in this rare outburst he set his wings stationary, circled, and dived back down.

At six o'clock the next morning I heard a fluttering commotion of excited "chips," and saw one of the young tumbling out of the nest. It caught the air with its wings, and awkwardly fluttered off into the woods. One of the parents was flying all around and with it, continuing to make excited "chip" calls. The other young had apparently already been launched, similarly. I found one of them perched on the ground under my truck. By afternoon all was quiet around the house—no more phoebe activity. However, the next day I found the whole brood of five stubby-tailed youngsters lined up on a dry twig under the leafy bough of an ironwood tree a short way into the woods.

Already at the next dawn one of the adults was chittering back at the old nest. It was expressing renewed interest in starting the season's second nesting cycle. Two days later the female was repairing and relining the nest to get it ready for her second clutch. Meanwhile, as she was incubating, her mate took on the responsibility of feeding the fledged young. These young fledged on 11 July.

In 2005 we moved to another house down the road. It reeked of cat piss, and it probably never had a phoebe; there was no phoebe ledge for a nest. My wife attended to replacing the rugs in the house and I attended to placing a phoebe ledge outside. I took three little pieces of board and nailed them each in a different place under the roof, to give the birds a choice of nest sites. As we hoped and expected, a phoebe pair did arrive during the first spring and inspected the nest sites I had provided. They chose the shelf next to the garage door.

Over the next two years these phoebes (probably two different pairs) made four attempts in that nest, but none of them was successful. Brown-headed cowbirds parasitized the nests with eggs, and then the newly hatched young were raided by chipmunks who had been attracted by the bird feeder, and who had somehow managed to climb the wall at the nesting spot.

After the successive failures of nests on the house I put up a shelf deep in the chicken shed, and it was quickly discovered. I was there when one of the birds found it; the bird cheeped and chattered excitedly, and I knew it liked what it had found in this new, very hidden, protected place. The pair soon produced five young there, but then one of the adults disappeared before they fledged. The other continued to feed them, but apparently not enough, because the cool wet weather that spring was not conducive to flies. One by one of the young died. Several fell out of the nest, as though trying to escape before starving. The next year, 2006, set records for rain, which hit just when the young were larger and needed much food. Again they starved, this time despite care by both parents, as did also the neighbors'. I removed the dead, and the birds then raised a second clutch in the same nest; this clutch fledged in early July. A second pair arrived then as well, and they laid a clutch of eggs in the ready-made but vulnerable and previously unsuccessful nest by the garage. Even before the clutch was finished (at the third egg) I heard the strained nuances of the adults' alarm calls. They attracted me, and as I suspected, the nest was once again empty, probably just raided by a chipmunk.

In 2007 the pair started refurbishing the same hidden nest in the chicken shed on 6 May. Egg laying was delayed because of cold and rain, but I eventually felt five eggs in the nest by reaching up and into it and feeling them with my fingertips. On 1 June I heard the adults announce the eggs' hatching, and the young fledged on 13 June. A cowbird was and had been around for most of this period, but I thought that the phoebes' nest had been safe. I did not expect a cowbird to come into the chicken shed for a nest tucked onto a ledge in a dark corner, so I had not even bothered to get a mirror to look into the nest. However, to my great surprise, in the days immediately after fledging I saw only one of the phoebe pair, and it was feeding a baby cowbird. A pair of cowbirds were then still around the premises. I never saw the baby phoebes, and I assume the male left with them while the female and "her" cowbird baby stayed.

The baby cowbird was still following the phoebe around, and persistently begging from her, until at least 30 June. However, at dawn on 18 June a loud singing refrain signaled that the male was back. I suspect he came back after having left the young on their own. Two days later I found the first egg (eventually there would be three) of the second clutch, again in the same nest. Even after the eggs were laid, the presumed female was still feeding the cowbird, and only occasionally incubating. It was not until 29 June that she was consistently incubating her eggs during the daytime. I suspect, therefore, that she was not free to incubate full-time until the male was back and able to feed her. Now the cowbird followed the male around. It looked as though the male was trying to escape from this nest parasite and its almost constant begging for more food. (A pair of phoebes reused the same nest the next spring. It again was parasitized with a cowbird egg.)

After the second brood fledges, our phoebes usually become almost silent. They still hang around the house and we occasionally see them until mid-September. After that we miss them, and we look forward with anticipation to seeing them the next summer. The phoebe connects me to the miracle of returning at all from I know not where, and to the mystery of I know not how.

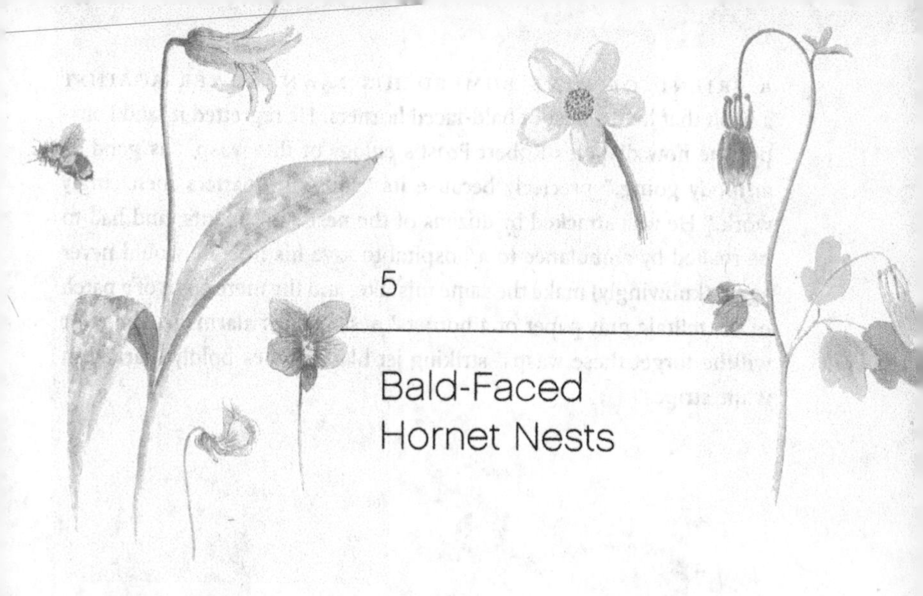

# 5

# Bald-Faced Hornet Nests

21 July 2006. I FOUND A NEST OF BALD-FACED HORNETS (*Dolichovespula maculate*). So far it's only about the size of a baseball. The hornets—or wasps—who were spooked out of the nest after I jiggled its branch inadvertently flew to attack and afterward went back to it, perched on the outside on the gray paper, and pointed in my direction. They rapidly vibrated their abdomens—they were shivering to keep up a high muscle temperature, necessary for fast flight and for more instant attack (nest defense). I advanced no farther. Luckily I had gotten only one sting. The wasps should have good hunting—it seems to be a good caterpillar year because there is a barely audible, gentle "rain" of caterpillar fecal pellets on the leaves at night. I also found a female black ichneumon wasp in the act of injecting an egg into a young tiger swallowtail butterfly larva on a chokecherry. I got the wasp and sketched the act, then watched the sapsucker station on the birch for an hour. Relative to all the activity last year there was so far not much action; but nevertheless three hummingbirds, two satyr or wood nymph butterflies (Satyridae), about a dozen bald-faced hornets, and a swarm of small flies (and in the evening, seen by flashlight, one flying squirrel) came to feed at the sap lick.

---

A FRIEND OF MINE BUMPED HIS LAWN MOWER AGAINST a bush that held a nest of bald-faced hornets. He regretted it, and I suspect he now disputes Robert Frost's eulogy of this wasp, "as good as anybody going," precisely because its "stinging quarters menacingly work." He was attacked by dozens of the nest's occupants, and had to be rushed by ambulance to a hospital to save his life. He would never again (knowingly) make the same mistake, and the mere sight of a patch of the telltale gray paper of a hornets' nest sets off alarms in him. Nor will he forget these wasps' striking jet black bodies boldly marked in white stripes.

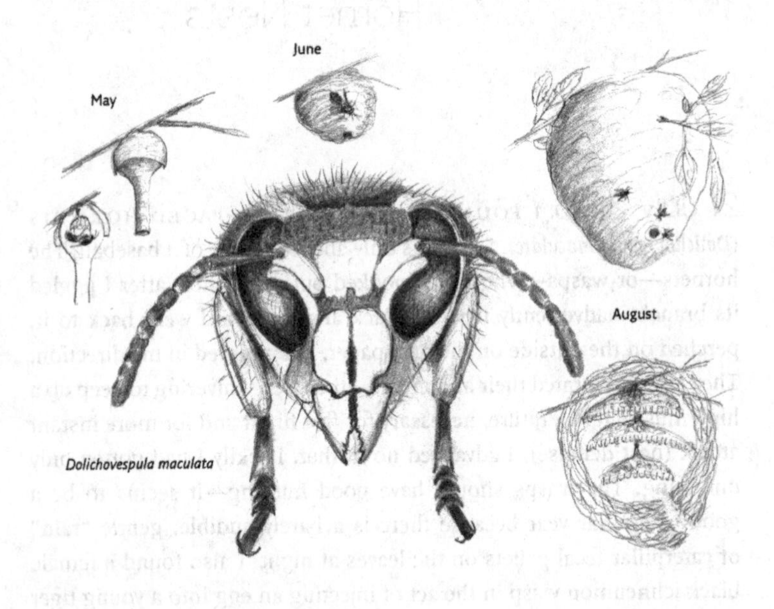


Fig. 12. Bald-faced hornet. Nests shown in the process of construction from golf-ball size, at left, to basketball size or larger at the end of the colony cycle, when it contains three or more vertical combs filled with eggs, larvae, and pupae.

A colony of bald-faced hornets starts out in the spring with one lone individual, the queen. She would have mated in the previous fall and then crawled underground to hibernate. As she comes out of her subterranean hiding place in the spring she shivers to bring her muscle temperature to above 95°F and then flies off in search of food. She hasn't eaten anything for at least seven months. And, like the ruby-throated hummingbirds who have just returned from Central America at the same time of year, she will most likely end up feeding on sweet sap alongside hummingbirds, where a yellow-bellied sapsucker has cut away birch bark to get at the tree's sap. She will feed at the sap lick, and later on in the colony cycle also from the sugary secretions with proteins that her own young regurgitate to her after she has fed them chewed-up caterpillars and other insects.

After she has refueled, the young queen becomes attracted to weathered dry wood, and with her mandibles working from side to side, she scrapes off swaths of fibers, mixes them with her saliva, and—presto—she has a gob of liquid paper pulp. She applies this pulp to the underside of a twig on a tree (or the bushes next to a lawn) to fashion a short stiff rod. Hauling in load after load of paper, she flanges the rod out to the sides and adds at the bottom of it a little battery of hexagonal cells (in shape much like the cups honeybees make out of wax, which they secrete from glands between their abdominal segments). She makes paper envelopes around her nursery of young, and while releasing a little sperm that she has stored from last fall, she also deposits an egg into each of the hexagonal paper cups. The offspring from these fertilized eggs are genetically females, though most of them will remain sterile "workers." Much later, near the end of the summer, she will lay virgin eggs, and these will become males. Afterward, at the appropriate time, she will again fertilize eggs from her stored sperm of the year before, and they will again produce females, but because of special treatment they will become fertile "queens" rather than mostly sterile workers (who may at times, however, produce sons, but never daughters).

Each larva hatches into a white grub in exactly the spot where the egg was deposited. Bald-faced hornets chew their prey into semiliquid gobs and carry these home to the nest to feed to their white grubs, who are neatly stacked in a nursery where the temperature is controlled to maintain their rapid growth. The nests trap heat, and the adults stay warm inside and remain ready for immediate, fast takeoff, even when the outside temperatures may be close to the freezing point. Throughout this time they indiscriminately defend themselves against potential predators, generally those who would make the contents of their nests—

the fat juicy grubs and the pupae that they turn into—a feeding bonanza. After achieving full growth in its respective crib, each larva spins a fine white cocoon around itself with silk from its salivary glands. This silk then seals the larva off in a coffinlike cocoon at the same spot where it started as an egg and then molted into a pupa before molting once more into an adult.

The queen's first successive batches of larvae need to grow fast and become a crowd of workers and nest defenders. She needs these sterile daughters, who are single-minded in their tasks and undistracted by sex, to help raise males and new queens in the fall at the end of the colony cycle. Keeping her brood warm accomplishes fast growth, and she encloses her brood comb with more and larger envelopes of insulating paper. She builds the paper layers from the top down, by attaching them to the petiole at the tops of the cells and extending it at the bottom edge until only a small entrance hole is left at the bottom. After she has enclosed her brood in her upside-down "bell" of paper, she also adds a long tubular extension, at the entrance hole, like that on a weaverbird nest. This construction reduces convective air movement through the entrance. Now, whenever she is inside the nest with her eggs, grubs, and pupae, she shivers and produces heat. The warm air is trapped in the bell chamber. She thus creates and maintains her own very local tropical "summer" climate despite what the weather may bring.

To the queen hornet, time and temperature are inextricably linked. She is programmed to behave in ways that shorten the development time of her offspring, so that she can raise as many as several hundred of them in the one summer allotted for her life. Whenever the nest cools, the eggs stop developing and the young stop growing. She needs to keep her baby factory going, to raise lots of workers during the summer. Her first batches of workers are slaves that promote her objective of raising many reproducing offspring later. These sterile workers cooperate because their evolutionary "objective" is to help to raise brothers and sisters who will insert the same genes they have into the next generation. Meanwhile, the colony as a whole is in competition with other colonies for resources to fuel its economy.

By middle to late October the shortening photoperiod has activated the enzymes that have transformed the leaves' physiology and caused them to be shed. The first heavy frosts are expected. On mornings when there is no wind, the remaining leaves on the tree come rustling down as the sun comes up, illuminates them, and melts the last thing that then holds them on the tree—a bit of ice. A week later all the leaves are down, and now many bird nests and wasp nests that were hidden earlier are suddenly revealed. There is little or no more insect prey and the wasps' colony cycle is completed. It is now safe to examine hornet nests to learn much about what they have accomplished during the summer.

The hornet nests are now the size and shape of a basketball, and they are sturdy structures that have withstood summer rains and produced generations of workers, then a brood of drones, and finally a brood of queens. I pulled one down (on 26 October 2003) to examine the details that would normally not have been accessible earlier, since an angry fleet of wasps would have come out like sidewinder missiles that seldom miss their mark, when one comes a-knocking. I assumed the nest would be empty. To my surprise, however, it contained twenty-eight new, live queens. All were comatose; they were too cold to fly, and they barely buzzed.

The hornet nest contained three horizontal combs that were hung one below the other, and this nursery was enclosed with multiple layers of insulating paper. The cells of the combs gave clues to the minimum wasp population the nest had produced. The oldest (topmost) comb had the smallest cells, which would have been the "cribs" of worker larvae (and their pupae). There were 240 of these, showing the silk tops of the cocoons where the young wasps had chewed their way out. The second two tiers of combs had larger cells; the first of these tiers had 212 cells that probably each produced a drone, and the lowest comb with the largest cells contained 257 large (queen) cells. The twenty-eight comatose queens still in the nest would have come from them, as well as the 220 additional queen cells. This comb had grown from a central point outward to the periphery, since the empty cells from which sixty-eight adults had emerged were in the center, surrounded by a ring of fifty-four now dead pupae (still in capped cocoons), in turn surrounded by 135 cells, also with dead larvae. The largest larvae were located toward the inside of the ring. There had been many casualties in this wasp colony—a loss of 46 percent of the total production of reproductives attempted; the summer had been too short for many wasps. However, since at least 212 drones and 40 new queens had successfully left the nest, it was probably not too short for the life of the colony.

A colony that faces a short summer, as in the arctic, has little option but to start making the drones and queens soon after the colony is founded and without building up a large worker force. However, in New England a colony could invest more time and energy in workers, the infrastructure of what Edward O. Wilson has called the "fortress factory," if there is assurance that it can reap the benefits of making good on the investment later on. Rearing many workers before switching over to the desired "product" is gambling that the weather will hold out. But a hornet (wasp) queen cannot be overambitious and push her luck too far in the summer, so that she ends up with a nest full of hundreds of larval young that will then all die when the cold sets in and the summer food runs out.

Of five other nests that I examined, four had switched over to make drones and queens in time to get all of them out of the nest to adulthood, but although they had many fewer casualties than the high producer, they also had less output of "product." On the other hand, the colony that waited the longest to build up its strength before switching to make its product lost 20 percent of its offspring, but it had the largest overall output—about 900 drones and queens—and this is what ultimately matters to wasps. Thus, its more modest "gambling" for a long summer season had paid off despite the loss of some offspring.

My interest in hornet nests is probably matched only by that of a bird, the red-eyed vireo. This bird uses the paper from hornet (wasp) nests to decorate its own nests. I have examined many dozens of vireo nests over the years and have seldom failed to find at least a patch or two of hornet paper conspicuously attached to the outside surface of each nest. Given what hornets can do, I felt that this bird has a strange taste for nest decoration, because the wasp paper was not serving as insulation on the birds' nests. Furthermore, the apparently obsessive behavior of putting one or more patches of flimsy wasp nest onto the outside of the birds' nest involves costs. Hornet paper is hard to come by. One can search for a hornet nest for many days and not find one, even in winter when such a nest is conspicuous from as much as 100 feet away. It must often be a burden for a vireo to find this material in the summer when the old wasp nests are not only rare but also nearly invisible, and when the new nests are vigorously defended. There must be an advantage to using the wasp paper that offsets the costs of getting it. I wondered if the wasp

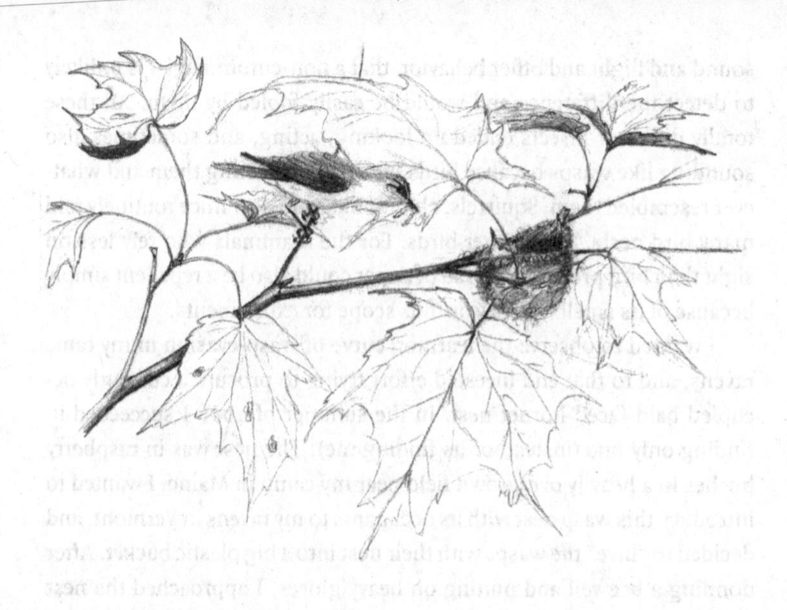

Fig. 13. Red-eyed vireo pair and their nest, which is typically decorated with several pieces of hornet paper.

paper, which might proximally serve as a decoration for the bird, might ultimately serve as a prop that fools potential predators into avoiding the vireo nest.

When vireos build their nests in late June the hornets' nests are still small—about the size of a baseball, the same as the vireos' nests. From a distance, from underneath, both look like gray blobs hanging from a twig. Might a crow, blue jay, chipmunk, or squirrel initially confuse one with the other? If these predators have experienced a wasp's nest defense before, then a mere glimpse of wasp paper on a shape that looks like a wasp nest may, as in some of us who have had an experience with these wasps, be sufficient to prevent them from coming nearer to make a close inspection.

Birds are quick learners. Like us, they stay clear of even a single wasp. And we know that this has been so for millions of years, because several different families of flies, as well as some moths and even some beetles, have members that mimic wasps so closely in shape, size, coloration, sound and flight and other behavior, that a non-entomologist is unlikely to detect the difference and would be easily fooled by them. All these totally different insects ended up looking, acting, and sometimes also sounding like wasps because birds have avoided eating them and whatever resembled them. Squirrels, chipmunks, and deer mice routinely raid many bird nests, as do other birds. For the mammals who rely less on sight than bird predators, a wad of paper could also be a repellent simply because of its smell. There was thus scope for experiments.

I wanted to observe the learning curve of wasp-evasion in my tame ravens, and to that end invested effort trying to procure a currently occupied bald-faced hornet nest. In the summer of 2006 I succeeded in finding only one (instead of its finding me). The nest was in raspberry bushes in a heavily overgrown field near my camp in Maine. I wanted to introduce this wasp nest with its occupants to my ravens in Vermont, and decided to "hive" the wasps with their nest into a big plastic bucket. After donning a bee veil and putting on heavy gloves, I approached the nest with bush cutters. The occupants shot out at me as soon as I wiggled the first raspberry twig. One immediately stung me on the wrist. I retreated to wait a bit for the pain to ease up, and after about five minutes I came back for another try. This time I had also brought my bee smoker, and I cranked it up until it was belching big gray puffs. Did the wasps retreat? Not in the least. I was surprised, since honeybees are almost instantly calmed by smoke. The wasps, in contrast, were not fazed. They came after me whenever I made even the slightest move. I retreated again, planning to come back at night.

This time I crept up slowly, lunged forward with a wad of toilet paper in my hand, and successfully plugged up their nest entrance hole before they had time to react. I then snipped the raspberry twigs holding the nest, dumped it into the plastic bucket, and sealed it, all in less than ten seconds. Within about that time the volume of the wasps' buzzing increased. The wad fell or was pushed out, and the hornets were exiting from their nest and hitting the inside wall of the bucket. It sounded as if someone was peppering it with dried peas.

During the five-hour drive from my cabin in Maine back to Vermont that night the wasps were undoubtedly jostled and agitated. But by the next morning the bucket was silent. I brought it into the aviary and cautiously lifted the cover, and as I did so only one wasp burst out to attack me. The rest were on the bottom, either barely crawling or dead.

My experiment was now a different one from what I had planned; it had morphed into trying to find out if ravens would attack an *undefended* bald-faced hornet nest. The results were clear. In minutes the ravens (who were naive with regard to wasp nests) tore the hornet nest to shreds and gorged themselves on the larvae and pupae that they found in the combs. I had achieved a positive result, as I hoped I would: wasp grubs are indeed prized bird food. Apparently wasp nests do require active defending, and in retrospect maybe it doesn't require an experiment to prove that a raven would back off a hornet nest when the defenders come streaming out to hit and sting with precision.

I saved the torn-apart wasp nest left over from the raiding ravens, and wanting to put it to good use, I wondered what our vireos (there is a pair near our house every summer) might accomplish with a superabundance of wasp paper. Might they plaster the whole of their nest with paper? I made three bundles of the paper by wrapping wire around it and hung them in the trees near the house. The vireo sang vigorously in the summer, but I found no nest until it was revealed after the leaves came down in November. I eagerly retrieved it, and to my surprise it had about the same amount of wasp paper as many of the other nests I had seen. Do vireos, after having incorporated a little bit of wasp paper into their nest, then find that their strange urge is satisfied?

In the next year our neighbor, while sitting on her porch in a thunderstorm in July, was suddenly attacked by a wasp that went for her eye. She suffered swelling down to her chest. After she located the nest I offered to help her get rid of it, and at night I went with a net I had made of wire mesh that was large enough for wasps to get through easily, but this mesh was covered with gauze that the wasps could not get through and that I could remove later. I scraped the rim of the net along the top of the nest to try to get it to drop into the net, and I had a piece of cardboard with me to slip over the net, in case the nest really did drop in, as I hoped. I would then, still at night, transfer the captured wasps in their nest into my (now nearby) raven aviary, where the wasps would settle down to rest. Later, still in the dark, I would remove the outside gauze, and in the morning the wasps would all be at their nest in the aviary.

The results showed that ravens respect wasps: the transfer of the wasp nest appeared to be a success; in the morning the wasps were flying into and out of the nest. No raven went near it. But gradually there were more wasps leaving the nest than coming back. After two days there were no more wasps at the nest, and the ravens then destroyed it and ate all the larvae and pupae with gusto.

Word then got out that I wanted wasp nests, and another neighbor offered a wasp nest that was under her porch. I tried the same approach, except that this was a much larger nest and it just barely fit into my net—and then, to my displeasure, I failed to get the cardboard seal over the top. In seconds after I dislodged the nest, dozens of wasps leaked out, and as I made a run for it my hands and arms felt as if they were on fire from the many stings. I flung the open net with the wasp nest into the back of my pickup, jumped into the cab, slammed the door shut, and spun off down the driveway. Later, back home in the dark, I maneuvered the nest into the aviary. As before, the wasps gradually left. After they had all cleared out, the ravens again destroyed the nest and ate the contents.

Although it would have been satisfying for me to get experimental results proving that the vireos' paper nest decorations indeed repel blue jays, chipmunks, red squirrels, and crows, negative results would not have proved that the origin of the paper is unrelated to that function. That's because the proximate results are not necessarily coupled to the ultimate results. It is possible that the paper on a vireo's nest now serves no useful function, and is more like our appendix, indicating to a previous function in ancestors. Interesting analogies with our cultural practices abound. We often behave in mysterious and enigmatic ways that no longer make sense but that originated from logical, functional antecedents—such as drinking white wine with fish, dipping a slice or two of raw onion into linseed oil to make a famous varnish, or adding a pinch of ammonium chloride to antirust paint, as Primo Levi illustrates in the chapter titled "Chromium" in his book *The Periodic Table.*

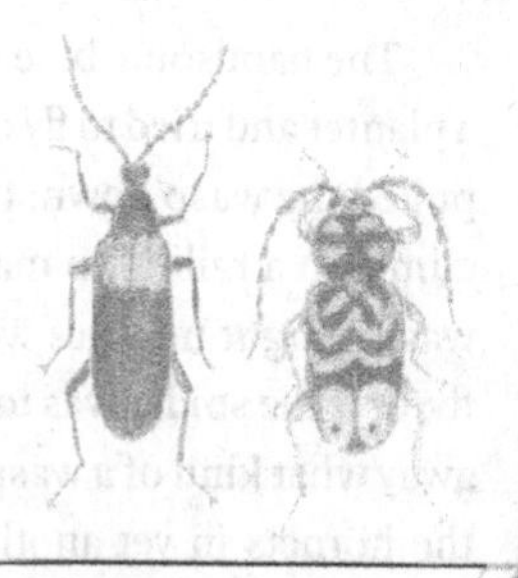

# 6

# Mud Daubers and Behavior

I'M NOT A GLUTTON FOR PUNISHMENT. IN ORDER TO get a kick out of nature—to enjoy insects, even stinging wasps—it is probably not necessary to risk life, limb, and anaphylactic shock. Indeed, I think I enhanced my educational experience, and possibly even advanced the cause of science, at least as much as I did by fooling around with the bald-faced hornets, during the summer of 2006, by sitting on our front porch sipping a glass of red wine with Rachel, my wife. It was early on a balmy August evening when big green darning needles, or dragonflies, were zigzagging back and forth in the clearing between our house and the bog. I was probably half dreaming about experiments I had done years earlier with a colleague, Timothy Casey, in which we had pinned them down and simulated overheating of their flight muscles (that they could normally experience in flight) by focusing a heat lamp on the thorax, and proved that they could stabilize their body temperature by shunting excess heat into their long cylindrical abdomen, which then served as a radiator. I had loved the control, the certainty, and the presumed cleverness of discovering a new phenomenon. Now, reminiscing, I still felt the glow, but in more ways than one, as I lifted the glass, looked out, saw the dragonflies, and then on the porch noticed a wasp dragging a spider.

The handsome blue wasp with dark wings carried the spider up onto a planter and tried to fly off with it, but the weight of the spider apparently pulled the wasp down: the wasp went only a short way before having to climb up a railing to make another short flight. I was wondering if the wasp's flight muscles were not hot enough to generate sufficient lift to fly, or if the spider was too heavy for the wasp to carry. I didn't know right away what kind of a wasp it was, except that it was a solitary wasp, unlike the hornets in yet another communal nest under our porch that I had recently offered, along with their nest, to my ravens.

As I was contemplating the wasp carrying the spider, Rachel casually mentioned having seen a wasp of the same description that had also been hovering along the side of the house, but she said that this one was carrying "a long piece of dry grass" and, furthermore, that she saw the wasp drag this grass into "a crack in the wall." I know that wasps do some amazing things. But they always do the same things—they don't vary their behavior. In short, I flat-out didn't believe Rachel. But maybe I should have. I will get to the reason why shortly.

A few days later, under similar circumstances, I again saw the dark blue wasp—a female (male wasps don't carry any objects)—carrying another spider. This time I followed her and saw her take the spider to a mud nest shaped like a long vertical tube that was plastered onto the south-facing wall of our house. Nearby, there were three of these nests of different lengths, neatly aligned adjacent to one another like the pipes of an organ. I knew from this nest that it was the organ-pipe mud dauber (*Trypoxylon politum*).

Insects are models that have given us a view into basic mechanisms of behavior and evolution. And birds, because they are emotional animals, provide a bridge to understanding ourselves. Their specialized behaviors—courting, vocalizing, nest building, foraging, habitat preferences, and strategies of parenting—are all deep-rooted in patterns that they are born with, as are those of insects. Insects show us how much can be done with a pinpoint-size brain, and they therefore seem magical. If so much can be programmed into such a small brain, how much more is possible with a brain like a bird's, which is hundreds or thousands of times larger? Nest building by wasps offers a direct comparison with birds, and when on that August day I found the organ-pipe (or "organ tube") mud dauber wasp carrying a spider to a mud nest on our house, I paused to ponder the differences between a bird and a wasp.

Eighty million years ago, in the Cretaceous, the birds' relatives, Maiasaurus, were hollowing out holes in six-foot scrapes to deposit their eggs and then care for their young in them. They nested in colonies, as many birds do now, probably because there is safety in numbers, and maybe also for defense. Undoubtedly, nest building by insects is at least twice as old.

Unlike dinosaurs, small birds had the option of building their nests in trees, and also of hiding the nests, which requires more finesse than digging a crude scrape. As with insects, each bird species makes a nest as distinctive as its plumage and just as circumscribed or encoded in its DNA. The goldfinch's nest is wedged into a vertical fork and is made of fine grasses and plant down. The oriole's is a bag made from the fiber of dead milkweed plants and is hung from the tip of a long limb on a spreading tree. The chestnut-sided warbler's nest is hidden near the ground in dense meadowsweet or raspberry vines and is a flimsy affair made entirely of very thin grass stalks. The tree swallow's nest in my bird box is made loosely from dried grass, and almost invariably the nests of this species are lined with feathers, preferably white ones. Robins build a hardened mud cup on leaves and debris and line it with thin grass strips. Wood thrushes commonly incorporate a snake skin into their nest. Catbirds line their nests with fine rootlets and use grape bark to garnish the exterior. Ravens and chickadees line their nests with fur. Often the specific items used in the construction seem to have little rhyme or reason, but the nests are always exquisitely "perfect" in functional design and constructed unerringly.

Many kinds of wasps make nests from clay or mud mixed with saliva, as barn and cliff swallows do: the mud hardens, and as long as it stays dry, it stays solid, like concrete. Like birds' nests, wasps' nests are shelters for their offspring. Solitary wasps, however, provision their nest not only with their eggs but also with food for the larvae after they hatch, and then seal the nest off to prevent parasites from entering. Some wasps, like the potter wasp, make a nest that closely resembles a narrow-necked jar.

The organ-pipe mud daubers I had been watching are another kind of solitary wasp that uses mud to make nests, but of a very different design

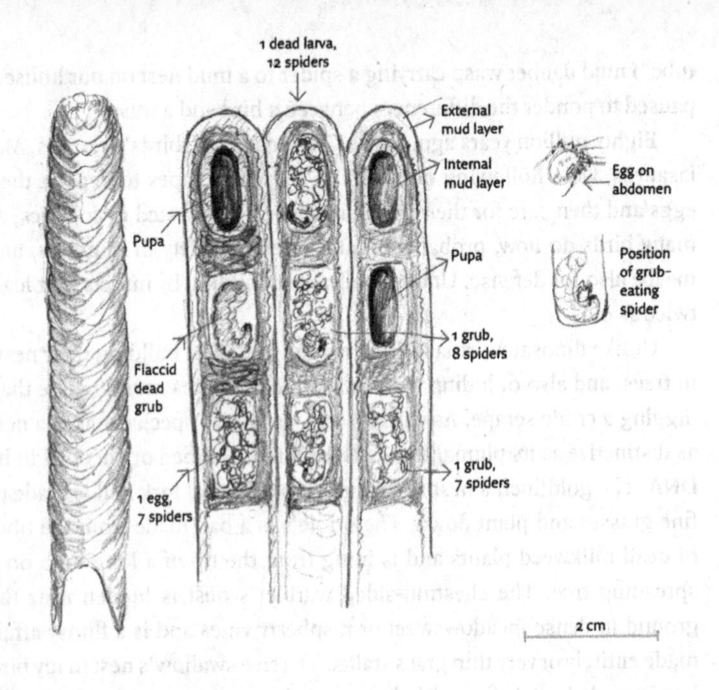


Fig. 14. Nests of the organ-pipe mud dauber. *Left:* Exterior of a nest, with ridges resulting from successive loads of hardened mortar. Three adjacent nests have been opened, each showing three cells containing various stages of development, from egg placed on fresh spiders to larvae ("grubs"), and to pupae.

from those of the potter wasp. The daubers fashion an upside-down tube with an entrance at the bottom. The tube is plastered against a wall (such as a cliff face or house wall). After making a first small section of tube, the female wasp (no male insects make or help make a nest, or provision one, or sting) collects spiders; jams the prey, still alive, up the tube; inserts an egg; and then makes a partition at the bottom so that the contents, the prey, won't fall out. She does not have to be concerned that the spiders will crawl out, because after she catches them she injects them with a chemical that keeps them in a zombie-like state of suspended animation. As a result, they don't struggle when carried, and they will still be alive and fresh when the larvae (which look like white grubs or maggots) need to feed on them days or weeks later.

The vertical tube design is efficient, because to have a cell for the next potential offspring, the wasp merely extends the bottom of the tube. And so she may continue to make one cell below another in a lengthening tube that may ultimately extend several feet down. As the summer progresses and the nest grows by being extended in a succession of cells at the bottom end, it eventually contains dozens of cells. The wasp lays only one egg in each cell. The cell that was made first (on top) may already have a pupa while the cell made most recently (at the bottom) is still being provisioned with spiders. An egg is deposited on the last spider put in each cell, just before the cell is sealed.

Using a sharp knife, I cut into the first nest and was surprised by how hard the walls were. This first nest I examined was divided vertically into only two cells. There were no spider remains in either one; the prey had already been eaten by the wasp larvae, down to the last leg. I cut into an adjacent nest, and here the lowest compartment contained eight fully intact spiders; these were of different sizes but were all similar and probably belonged to the same species. As I pulled them out, one by one, they wiggled only slightly. Each raised its front legs briefly in a defensive gesture, then quickly let them droop. However, as long as I didn't touch them they remained motionless. The largest of the eight spiders had a yellow oblong egg attached to its side. The compartment above this one (sealed off from the one below) also contained eight live spiders of the same type. Only one was dead—it had a collapsed, flaccid abdomen with fluid leaking out, a sign that the wasp larva was eating it. One compartment, the upper one, contained a pill-like pupa; the lower compartment had a dead larva that was as flaccid as the spider being eaten, so I assumed it was diseased and dead.

The following summer, however, I learned that the flaccid, seemingly dead larva probably hadn't been "diseased" at all. I had saved several nests to rear out the wasps. Three wasps emerged, and to my great surprise they were not organ-pipe mud daubers. Instead, they were scoliid wasps. Scoliids are well-known parasites of scarab beetle larvae, such as those of June beetles that live in the ground; apparently here was a species of wasp that parasitizes a mud-nest-dwelling larva of another wasp.

Although only the females build and provision the nest, the male organ-pipe mud dauber is one of a very few that stay around during the nesting. He is thought to help guard the nest against potential intruders while the female is away hunting prey for feeding the young (when the nest is left open). Apparently males had been absent at the parasitized nests I had found, or else they had been negligent. In this case the male's apparent negligence resulted in the death of his and the female's offspring. However, in other instances, when flies enter the nest to lay eggs that feed on the wasp's prey, the larva's food, the offspring are not necessary killed. Instead, they grow into miniatures because of food deprivation (O'Neil et al. 2007).

My observations of nature from our front porch soon led to other, even more startling surprises. You may have guessed it—a few days later, as we were again sitting on the front porch drinking our usual after-supper glass of red wine in the gathering dusk, I thought I saw a light-colored piece of straw about half a foot long "flying" horizontally and then hovering in midair. That caught my attention—I looked closer and saw a black wasp that seemed identical in form to a mud dauber, and it was carrying an object. I jumped up in my excitement, and the wasp was spooked and flew off. The proof eluded me, but it dropped its "prey" onto the porch. I picked it up—definitely a long piece of dry grass!

Expecting the wasp to return, I waited. After about twenty minutes it did return, carrying another piece of grass. This time I was ready with an insect catcher net, and I snagged the wasp along with the grass it carried. The wasp was about 0.6 inch long, and the blade it was carrying was about 2.4 inches long. The wasp looked superficially almost identical to the mud dauber, but its body was black rather than black-blue and its wings were smoky-colored instead of blue-black like the mud dauber's. It was a different species of wasp, which I later identified as *Isodontia mexicana*. I also learned later that rather than making clay organ-pipe cavities for its nest and filling them with spiders, this wasp lines preexisting cavities with grass and fills them with paralyzed crickets or katydids.

The next year the organ-pipe mud daubers were gone from our house. The remaining nests on our house had all been pecked open, probably by woodpeckers or chickadees. I checked in a neighbor's barn, and deep inside, above the horse stable in the ceiling of the haymow, I found numerous mud nests of another species, the blue mud dauber, *Chalybion californicum*, plastered onto the wooden beams. This species attaches successive cells laterally, next to each other, rather than underneath each other as the organ-pipe dauber does. These nests were also packed with paralyzed spiders, although any one cell contained a variety of species. There were white and yellow crab spiders, brown orb web spiders, and still others. In the western United States this species of mud dauber is renowned for preying on the infamous black widow spider.

Fig. 15. A three-celled nest (detached from a barn beam) of the blue mud dauber, and the spiders out of one of these cells along with the wasp grub found with them.

The summer work of the insects was, as it turned out, not totally benign; one managed to "bug" our heating system. When I went to check out and prepare our outdoor wood-burning furnace, which pipes the heat into our house through a water system, I found it faulty. The water gauge, a vertical tube, stayed empty rather than filling up as it was supposed to when I turned on a valve, so I could not risk making a fire. Luckily I noticed some debris in the bottom of the tube and thought of leaf-cutter bees. These solitary bees make their nests in the galleries in wood excavated by long-horned beetle larvae, or substitutes thereof, and line them with green leaf pieces that they cut from whole leaves. The green leaf then wraps their egg and the pollen for the larva that they add next before sealing the tube nest with mud. I sent some of the remains that I could fish out of the furnace water gauge to Kevin O'Neil, an expert on bees, and he confirmed that a leaf-cutter bee had indeed bugged our furnace.

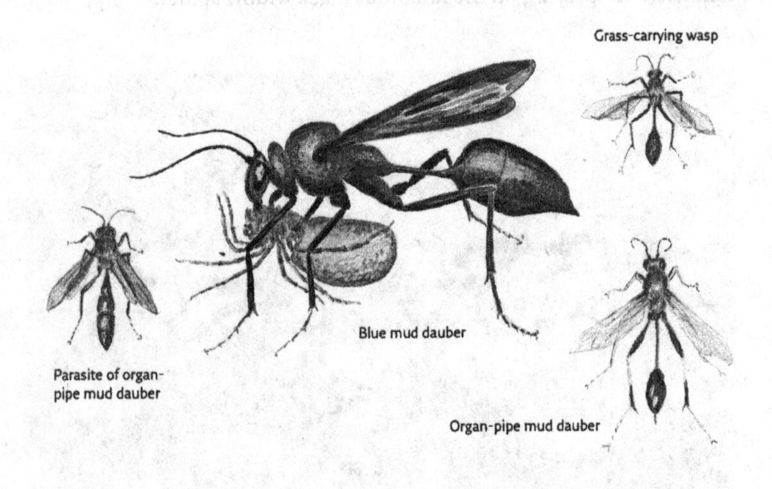


Fig. 16. The four wasps (not to scale). *Center, enlarged:* Blue mud dauber carrying a paralyzed spider. *Right:* Grass-carrying wasp and organ-pipe mud dauber. *Left:* Nest parasite of the organ-pipe mud dauber.

These four species of "house-barn" wasps, three of which are anatomically similar, plus the bee, have behaviors that are strikingly different. Their behaviors are, like their anatomy, presumably encoded on their DNA but expressed only in response to specific cues. Like most adult insects' lives, theirs are short—a matter of weeks. Since there is little time for experimenting and learning, such animals must necessarily get everything right almost from the moment that they emerge from their pupa. The newly emerged mud dauber flies off soon after its wings harden, and it finds mud! The dauber "knows" how to pick up and carry mud, where to go with it, and—more amazingly—how to make a nest out of it, of a very specific shape. Of the hundreds of thousands of potential things the wasp could search out, it then looks for specific kinds of prey (or it looks in specific ways so that it finds only those kinds). It responds to very specific, minute details out of thousands that it encounters. Just as we pick out what vegetables to buy at a market, it makes an incredible number of choices every second. Its choices are predetermined by genetic instructions. There is little flexibility. And the next species has entirely different instructions. If I wanted someone to exactly duplicate any wasp's behavior, I would have to write an encyclopedia of instructions, and even so, he or she would almost invariably make innumerable mistakes. But wasps, with a brain that is smaller than a pinhead, accomplish their specific tasks perfectly and without practice, flying out into a world they have never seen before. Aside from the mystery of how wasps can do so much with so little, there is the mystery of how what they know is passed faultlessly from one generation to the next.

Mutants of "motor" behavior (such as stepping, flight, climbing, walking, and reaching over gaps), learning, and memory are known, proving that behavior is indeed tied to genes. But how we get from protein gene products to *programs* of behavior consisting of hundreds of precise choices and actions—as opposed to behavioral tendencies—remains one of biology's great mysteries. And here I met it head-on, right in front of my nose on our front porch.

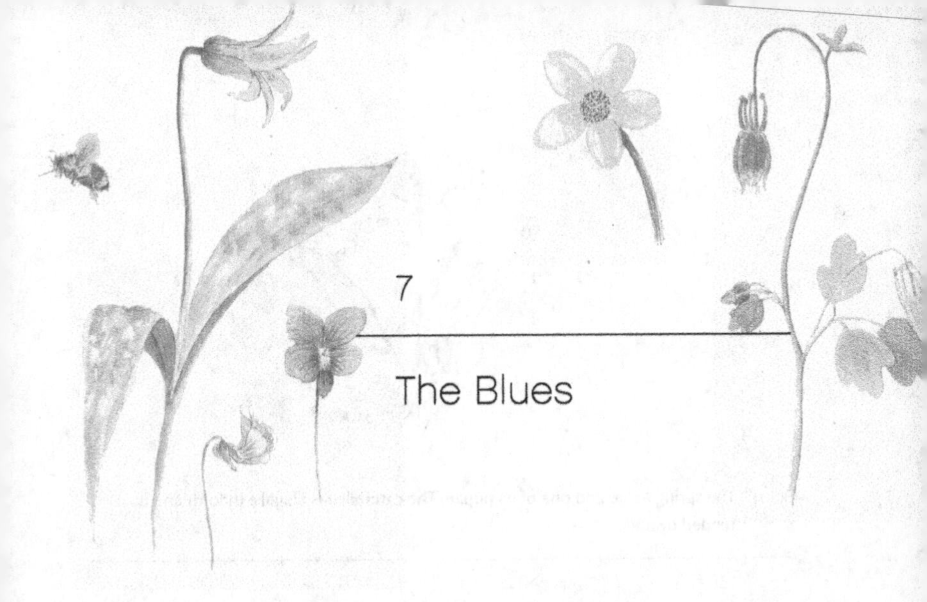

# 7

# The Blues

IT'S 12 MAY AND I'M AT CAMP IN MAINE. MR. WITHAM, who lives alone in a shack at the bottom of the hill, tells me that the ice was still on nearby Hills Pond last Friday. But today I heard the loons from that direction, so there is probably finally some open water. There is still snow along the roadsides, although the recent warm days have greened the aspens on some of the hillsides. Leaf-eating caterpillars are already on the trees, because swarms of warblers, vireos, and red-breasted grosbeaks arrived, almost to the day, as soon as the trees unfurled their leaves. It is warm and sunny enough today for the red Formicidae ants that live in and around my cabin to be active as well. They were running in a column up and down the trunk of the birch tree next to the cabin. A couple of the ants in the column were dragging a small caterpillar. I took it from them, and identified it as a species of "prominent" (notodontid) larva that I had in previous years found feeding on birch leaves. I then returned it to the base of the tree, and within seconds four or five ants almost pounced on it. But my most memorable sighting today was a tiny butterfly, *Lycaenopsis argiolus*, the spring azure. This butterfly is not uncommon when the poplars leaf out, and I may see more than one on any walk along the path up to my camp. The spring azure's common name is apt, because this is the first butterfly to emerge from pupae that hibernated all winter (some, like the mourning cloak

Fig. 17. The spring azure and one of its pupae. The caterpillar is sluglike in form and is tended by ants.

butterfly, overwinter as adults). It is hard not to be entranced by this butterfly. The surfaces of its sky blue upper wing glint like mirrors of the sky as it flutters over last year's pastel-colored dead vegetation seeking the first spring flowers, often while there are still patches of snow on the ground. When the azure flies, there has been warm weather before it, and summer is not far behind.

The spring azures' green sluglike caterpillars feed on buds and flowers of violets, and they are usually "tended" by ants. Ants kill most other caterpillars with little hesitation, yet they do not eat these caterpillars. Instead, they associate with them and repel predators as well as insect parasitoids, effectively acting as the caterpillars' bodyguards. The secret of the caterpillars' allure to the ants is that the caterpillars exude droplets of a sweet, protein-rich nutrient broth from glands on their backs when palpated by the ants' antennae, and the ants lap this broth up.

MOST MEMBERS OF THE AZURES' LARGE AND VERY INTERESTING family of butterflies, loosely called "blues" (although not all of them are blue), have small sluglike caterpillars. Few people except experts and aficionados (most famously including the novelist Vladimir Nabokov), who have studied them sufficiently to be able to distinguish them and name new species, know where to find them. But ants and these caterpillars find each other.

Many blues caterpillars are renowned for their close association with ants, and unlike the vast majority of caterpillars, some of these feed ants as the azure does, others move into the ant nests and are there fed by ants, and still others move in with the sheltering ants and are carnivores of their young. One thing has led to another in a likely evolutionary progression.

Once an organism has "found" or invented a successful strategy, the stage is set for replication, amplification, and modification. The great species diversity of the blues probably arose because an ancient ancestor had hit on the ant guard, thereby creating a new, safer niche for itself—one less exposed to birds and other predators as well as parasitoids. It is not likely that thousands of species of blues each independently discovered ants, but they have done different things with ants, and vice versa. Some of the most intricate and interesting associations are found in the nearly perpetual summer of the tropics.

One of the species most highly differentiated from its ancestral condition (evolved) is the moth butterfly, *Liphyra brassolis*, a huge Asian and Australian "blue" with a wingspread of nearly three inches. It is a relative of all other blues, but it is colored brown and black. In this butterfly the larvae make the ultimate use of ants: they are protected by living inside the nests of a very aggressive tree ant, *Oecophylla smaragdina*, and then they eat the ants.

These ants' nests are made by cooperation between the young and the adults. The larvae produce silk thread from their salivary glands, and an adult holds a larva in its mandibles and waves it back and forth between two leaf edges. As the larva exudes its sticky silk, it attaches to the leaves, which become glued together to produce the shelter that is the nest—the shelter that the butterfly larvae then use as well. But how do the butterfly larvae gain entrance to the ants' citadel? What we see now is a result of a long evolutionary arms race, a contest that these caterpillars have apparently won, because they get all the benefits whereas the ants gain nothing. Presumably, as with the spring azure in Maine, the caterpillars were originally symbiotically or at least benignly associated with ants.

The exact steps from there to here are obscure, but details of life histories suggest the problems faced and the possible solutions. At present the moth butterflies lay their eggs on the underside of branches, where they are less likely to be detected, and then the young caterpillars crawl into the leaf nest on the twig high in the tree. Undoubtedly they suffered casualties for millions of years, but over that time they evolved a thick leathery skin that ultimately became a nearly impenetrable armor and transformed the originally sluglike form of the blue's caterpillars into a little tank. The armored caterpillars are equipped with "treads" that allow them to attach themselves to the ant's nest substrate, primarily the leaf surfaces, so that the ants can't turn them over and bite into their soft underbellies, and can't detach them to throw them out.

The problem comes when the caterpillar must molt to the pupal stage, since the fresh pupal skin is necessarily soft, thin, and easily penetrable. However, these caterpillars have solved that problem, too. When they molt to the pupa they remain inside the caterpillar-tank skin, rather than discarding the skin as all other caterpillars do. But staying encased in the armor could be a problem later on, when the adult needs to emerge. To solve the problem of getting the tough armor off, the caterpillar shell has built into it predetermined lines of weakness that allow the soft, emerging butterfly to more easily crack out of the tank. However, the adult butterfly emerging within the ant nest is still necessarily soft; otherwise, it could not expand or inflate its wings, and it could then potentially be dispatched by the ants. Again, the butterfly has a special solution—unlike other butterflies, this species is covered with a dense layer of white, mothlike, powdery scales (hence the name "moth butterfly") that gum up the mouthparts of any ants trying to bite it. The loose scales stay on long enough for the soft, still vulnerable butterfly to escape the ant nest and then inflate its wings and harden its cuticle outside the nest. The rest of the protective powdery scales are eventually sloughed off on their own.

The *Oecophylla* ants that host these caterpillars are very aggressive, and so they are all the more useful to any caterpillars that can breach the ants' defenses. These day-active ants are also aggressive toward other ant species and try to evict them from their trees (kill them) whenever possible. Yet another ant species, *Polyrhachis queenslandica*, may live on the same tree with the *Oecophylla*. It avoids overlapping by being strictly nocturnal. In the daytime, when the *Oecophylla* are active, the *Polyrhachis* avoid being killed by staying in their nests—a couple of superimposed leaves glued together and sealed along the sides, with only two narrow tubelike entrances built into opposite ends of the nest. During the daytime, guard ants position themselves at these entrances and neatly plug them with their flat heads. No *Oecophylla* can get past these head plugs to enter, nor can any caterpillars. Nevertheless, these ants also host a blue caterpillar—one that has evolved a strategy exactly opposite that of the moth butterfly.

This blue, *Arhopala wildei*, lays its eggs directly on the *Polyrhachis queenslandica* ant nests or on twigs near one of the two nest entrances. To these ants, the caterpillars are not enemies; but their *Oecophylla* enemies on the same tree would eat them. To avoid that fate, these blues' eggs hatch at night when the *Oecophylla* sleep, and the nocturnal *Polyrhachis* ants then safely carry the young caterpillars into their nest before dawn. Once inside they treat these blues' caterpillars as their own larvae, and when the caterpillars molt the ants even assist them by pulling apart the cuticle to help them emerge, in the same way that they help their own larvae to molt. And if the nest is disturbed the ants carry the caterpillars to safety along with their own brood. The bribe? The *A. wildei* caterpillars have a gland on their posterior end that produces an ambrosial (to the ants) exudate. The caterpillars offer this tasty treat by raising their rear end to allow direct access to it whenever an ant approaches them. Apparently they have a scent that mimics that of the ant larvae so that the caterpillar is confused with one of them. Meanwhile, the treat ultimately comes from the ants themselves, because the caterpillars gorge themselves on the ants' eggs, larvae, and pupae.

There is so far no explanation why these blues' life cycles are so much more complex than that of the familiar azure that ushers in the short summer in Maine. But perhaps in a perpetual summer, such as a tropical one, there is more time and opportunity to evolve complex social relationships.

By far the majority of the larvae of the world's butterflies and moths feed on leaves, often the plentifully available leaves of trees, and one might not expect that much sophistication and intrigue might be involved in the behavior of harvesting them. But, as the blues indicate, one should never underestimate even a caterpillar!

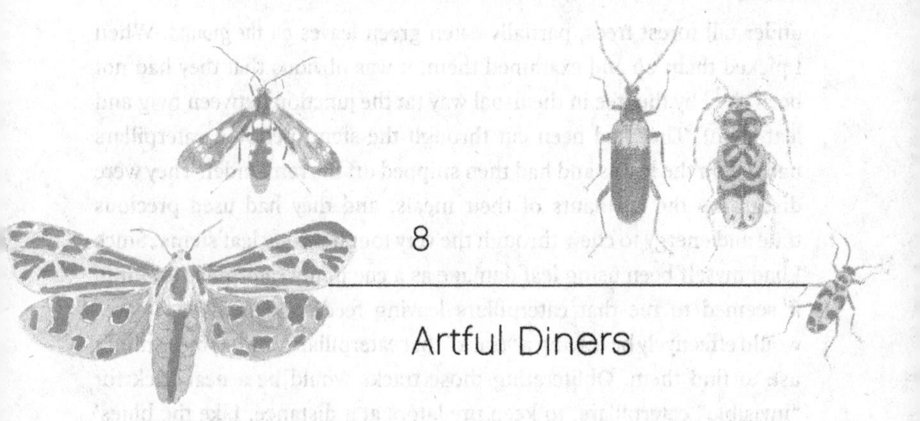

# 8

# Artful Diners

2 August 2006. TEMPERATURES ARE NOW REGULARLY IN the high eighties and low nineties, and as usual, the air is muggy and sweltering. The cicadas' buzz is now the definitive summer sound. The house wrens are totally silent—a big change! The eggs of this clutch (their last) are about to hatch. Do they already know this? Is that why the male now no longer sings? In the trees I see a wiggle of leaves here and there, as birds search for caterpillars. I saw the first roadkills of the large caterpillars—mostly sphingids—about ten days ago; they are leaving their food trees to wander before pupating underground. Monarchs float by lazily, but they occasionally accelerate in a few wing beats. One of them flies in a more steady, stately manner, and then lands on a bush. It spreads out its wings, and I notice that another one (which has its wings closed) dangles from it, attached by its genitals. Extreme mate-guarding. Maybe I will soon see more of their white, yellow, black-striped caterpillars on the milkweeds. Viceroy butterflies, the monarch's mimics, are making their first appearance. At night we see the flashes of distant lightning storms, and occasionally hear rumbles of thunder.

ONCE, WHILE STILL SEARCHING FOR CATERPILLARS, I SAW something that made my eyes pop. It was in midsummer, and I found, under tall forest trees, partially eaten green leaves *on the ground*. When I picked them up and examined them, it was obvious that they had not been shed by the tree in the usual way (at the junction between twig and leaf stem). They had been cut through the stem (petiole); caterpillars had fed on the leaves and had then snipped off the remainder. They were discarding the remnants of their meals, and they had used precious time and energy to chew through the very tough woody leaf stems. Since I had myself been using leaf damage as a cue in my caterpillar hunting, it seemed to me that caterpillars leaving feeding damage on leaves would effectively be leaving a "track" that caterpillar-hunting birds might use to find them. Obliterating those tracks would be a neat trick for "invisible" caterpillars, to keep predators at a distance. Like the blues' caterpillars and their ants, the story of leaf-eating caterpillars and birds also involves an evolutionary arms race, and it is waged all day every day all summer long.

Birds have specialized behavior to capture insects, and insects have specialized behavior to try to avoid capture. Throughout evolutionary history, the field of battle in this arms race keeps shifting as each participant keeps up with the other. Those that don't keep up will cease to exist. At no time is the race more intense than after the birds' young hatch, when the parents' real work of caterpillar hunting begins.

Most small northern forest birds attempt to raise four to six young in any one clutch. It requires an enormous daily investment in foraging to feed that many babies and bring them to adult body weight in about a week. To predators, baby birds are helpless gobs of tender meat whose noisy competition among themselves for their parents' attention—in order to be fed—is nothing more than a convenient advertisement of where to find them. Thus the fledgling stage is the most dangerous time during a bird's life, and there is a huge advantage in becoming capable of flight and getting out of the nest as soon as possible. To maintain the offspring's phenomenal, sustained growth spurt, the adults must feed their youngsters every few minutes, and the food must be easily digestible protein. For most forest birds, that means caterpillars.

Much of what applies to baby birds also applies to caterpillars, of course, except that they necessarily feed on foliage, a decidedly low-protein diet. Caterpillars have few hard parts—no skeleton and usually no "fur." They are easy to digest, and many of them require no preparation before eating: they can be swallowed as is. Like baby birds, they also need to grow fast, but because most (though not all) of them have a vegetarian diet, they don't reach their full weight nearly as fast as baby birds. For many caterpillars survival requires a delicate balancing act—hiding versus feeding. This is a hard compromise, because the tree's leaves are necessarily exposed to sunlight, where it is difficult to hide.

Birds, wasps, and flies have been preying on or parasitizing caterpillars (or both) for probably at least 100 million years. Year in and year out, the great majority of hatchlings of any one clutch of moth or butterfly eggs, consisting of perhaps two hundred, will be eaten. Such relentless pruning has necessarily left a deep imprint on the caterpillars.

Caterpillars are startlingly, dazzlingly diverse in their shapes, colors, and behavior. Some are protected by potent poisons. Others have greatly reduced their palatability by developing woolly hairs or sharp bristles. All the caterpillars that are conspicuous to us and to birds are relatively immune to wasp and bird predation; those that are hard to find tend to be the most prized as food by birds. Not surprisingly, therefore, the majority of edible caterpillars use a variety of strategies to keep themselves well hidden, if not to make themselves nearly invisible, and these are the birds' main summer fodder. It might seem that if more than 90 percent of any clutch of moth or butterfly caterpillars will get eaten, then they are not well adapted to evade bird predation. But as one member in an evolutionary arms race gets better at hiding, the other gets better at finding. Birds are very good at finding.

I remember my own first finding of a caterpillar as an exquisite experience. I was in elementary school and was picking berries in the woods. I was startled to discover among the raspberries something very beautiful—a plump green body decorated with red tubercules sprouting short black bristles—and I have been enamored of caterpillars ever since. As a graduate student I chose to study how the tobacco hornworm caterpillar is programmed to handle leaves without moving from its attachment point at the base of the leaf, and without leaving any scraps. Avoiding being eaten is usually an even greater problem for a caterpillar than finding enough food. To discover how caterpillars might escape detection by birds, I first used students as stand-ins for the birds.

I got reacquainted with caterpillars during a summer in the late 1970s at the University of Minnesota Field Station at Lake Itasca, where I helped teach a field ecology course. Each of the three instructors designed "field projects" for our select group of about a dozen graduate students in biology. These field projects had to involve the local flora and fauna, and I was spending a couple of days getting oriented in the local woods, looking for potential projects. It was then that I found the clipped-off partially eaten leaf of a basswood tree.

In my previous fieldwork with bumblebees (in Maine) I had studied individual bees and found that they became specialists, developing individual skills for finding and working on specific kinds of flowers. In any one field with several kinds of flowers, one bumblebee might, for example, search out the clover flowers while ignoring most of the goldenrod. Meanwhile, another in the same field might visit the goldenrod and ignore the clover. The bees maintained their specialties independent of the abundance of other flowers. They developed "search images" of what flowers to look for. Birds (and humans) presumably also use search images to help them find specific caterpillars. It helps to know what you are looking for, but this knowledge usually comes at the cost of not noticing the remainder.

To demonstrate the significance of the search image to my students, I "planted" four stick-mimicking caterpillars (Geometridae) on a poplar sapling, and then gathered the students around it and asked them to search. I told them that there were four caterpillars directly in front of them, but gave no clue of what these looked like. (Two of the geometrids mimicked live green twigs, and two mimicked dead brown twigs.) No "successful foragers" were allowed to give clues to anyone else about what they had found or where they had found it.

I had expected that it might take seconds, or at the most a minute or two, to find any one of the caterpillars, which were only a few inches in front of our eyes. I was indeed surprised to discover that despite their earnest and continuous searching, few of these naive but eager hunters discovered a caterpillar within the first half hour. But those who did eventually find one then located the other, similar one in a minute or less. That is, as predicted, after the students knew what to look for, everyone's performance increased enormously. This generalization has large implications. If birds search for caterpillars similarly to bumblebees and students, then there is a great advantage for each caterpillar species in having a different disguise from the next: that is, one not included in the existing repertoire of the predator's search images. This implies that it helps to be rare and different.

There is an enormous variety of caterpillars among the 250,000 species of moths and butterflies. Some species look like leaves or parts of leaves; others resemble twigs, bird droppings, or debris; some blend into the background of bark that they choose to rest on when not feeding; others cover themselves with debris taken from their background. In a slide demonstration of caterpillars that I have given several times, I take the audience on a virtual caterpillar hunt featuring various showy caterpillars perched on their food plants. After a number of practice slides, to get the audience used to finding caterpillars in the projected images, I give a test: I show a picture with real hidden caterpillars, or things that look like them, or both. Invariably even professional entomologists have been fooled, either not seeing the real caterpillars that were in plain sight and magnified 1,000 times on the screen, or pointing out something they wrongly assumed to be a caterpillar. My own skill at finding caterpillars despite their various tricks involved looking for fresh feeding damage on leaves, in order to focus the search on a smaller area, since most (though not all) caterpillars don't move far.

After I had found the first clipped-off partially eaten leaves in Minnesota, I searched and eventually found the inconspicuous stubs of leaf petioles still attached to a twig of the tree where the leaves had come from. The twig with its many remaining leaves looked un-grazed, and I would normally have passed by without a second look. But now I looked closely, and as expected found a large caterpillar (a big brown *Catocala* moth larva that was nearly invisible as it rested tightly pressed onto a nearby branch, mimicking the irregularities of the bark). I later watched and photographed the caterpillar and learned that it spent all day motionless in its disguise on the branch. In the evening it moved quickly out onto the branch, fed on a leaf that was far too large to consume in one meal, and after eating a portion of this leaf, backed down the petiole and chewed through it so that the leaf remnant dropped off. The caterpillar then walked away and resumed its position in its hiding place on a twig. I observed and photographed similar behavior in many species of caterpillars, but these were only those species that are "invisible" and hence have evolved to avoid predators that hunt by vision instead of scent.

These caterpillars were also the same ones that, while still feeding and before clipping, pared the leaves down so that they looked smaller rather than conspicuous because of tatters or holes. Some of these species, like the prominents (Heterocampidae), disguised their feeding damage by fitting their own bodies into the area at the leaf edge that they had consumed, and their bodies mimicked the leaf edge they had removed even to the extent of having fake leaf blemishes and leaf-edge patterns resembling those of the leaves of the tree they fed on.

Only those caterpillars that were routinely predated by birds had cryptic body markings, practiced leaf paring, held specific feeding positions on the leaf, and engaged in leaf clipping. The bristly or brightly marked ones that were not eaten by birds (but were still parasitized by wasps and flies) were "messy" feeders who did not clip leaves. These observations suggest that leaf clipping is part of a behavioral repertoire in the game of hide-and-seek with birds.

Although it seemed like a safe bet that birds hunt for the palatable "invisible" caterpillars, as I did, using leaf damage as a cue, nothing can be taken for granted. Any such idea needs to be tested, generally through long, tedious work that may take months or years to carry out, and that almost invariably leads to surprises. Here was a project that needed to be done, and I invited a friend and colleague, Scott Collins, to join me at my cabin in Maine for a summer's work and fun, to do the critical tests and determine whether a bird might be capable of learning to hunt for caterpillars by using leaf damage as a tracking cue.

Scott and I decided to use chickadees as our subjects. They were abundant, tame, and easy to catch in mist nets we set out in the woods. We erected a screen-cloth aviary in a clearing we made in dense woods where chickadees would feel at home, and in that enclosure we set up two compartments: one was to hold our birds (an eventual six), and in the other we installed ten small birch or cherry trees every other day. Our results were clear—all of our subjects quickly learned to search preferentially on trees (whose locations we kept switching around) with leaf holes, if such trees had yielded food before. In subsequent tests we also determined that they could be trained to search at specific kinds of trees (such as birch versus maple or cherry), and also to use real caterpillar feeding damage as opposed to our experimentally damaged leaves, which had holes made with a paper punch.

Birds in the wild hunting for caterpillars face a considerably more difficult problem than our chickadees faced in our enclosure. If the natural situation in the surrounding woods were as simple as we had to make ours in order to answer one specific question, then birds would probably be behaviorally hardwired to be attracted to caterpillar feeding damage. They aren't. Our chickadees *learned* to associate leaf damage on specific kinds of trees with food. However, to be indiscriminately attracted to feeding damage, as such, even on the proper species of tree, could be a liability in the field, because trees accumulate much leaf damage throughout the summer (or over a period of up to six years in the tropics), and that damage eventually ceases to be a clue to whether or not a caterpillar is still in residence. Early in the summer, when all the leaves are fresh, leaf damage can suggest that a caterpillar fed recently nearby; but in the fall a damaged leaf could mean that a caterpillar was there three months ago.

There is a second potential problem with using leaf damage as a tracking cue: the caterpillars that are the least palatable leave the most leaf damage. As already mentioned, caterpillars that are bristly or spiny (or both) and caterpillars that are poisonous, which are *not* routinely eaten by birds, are "messy" feeders—they make no attempt to hide their feeding tracks. They feed on the softest leaf tissues and often leave the tough leaf veins and the rest of the leaf hanging. Using leaf damage, as such, could therefore be a very misleading clue in hunting for palatable caterpillars. The contrasting behavior of the specific caterpillars that are less favored by birds thus provides independent evidence that parasitoids are probably not searching primarily by visual cuing on leaf damage.

I could distinguish whether a leaf had been fed on by a palatable or an unpalatable caterpillar, since unpalatable caterpillars ate a leaf into tatters, and palatable ones pared it down gradually to reduce tatters. I wondered if birds, to whom being able to make the distinction would matter, could also learn to differentiate leaves eaten by palatable versus unpalatable caterpillars. I talked about this problem with the animal psychologist Alan Kamil, who had recently used blue jays in lab studies to test these birds' visual acuity in discrimination in finding cryptic moths. In his lab, jays undergo individual choice tests, after they are trained to peck a screen in response to specific pictures projected onto it. They get a food reward if they respond to the "correct" picture. I sent his lab a series of photographs of leaves eaten by palatable versus unpalatable caterpillars, and Pamela Real et al. conducted the experiments. To my satisfaction, but not to my great surprise, they reported that "the jays exhibited little or no difficulty" in distinguishing the pictures of leaves that these two kinds of caterpillars had fed from. Not only that: they generalized. The lab jays learned to peck at pictures on a screen of a leaf partially eaten by a palatable caterpillar, and to ignore pictures showing unchewed leaves or leaves fed on by unpalatable caterpillars.

29 MAY 1985. I WAS WALKING UP THE PATH TO MY CAMP IN Maine. The aspen leaves had unfurled pea-green leaves about a week earlier, and as I walked under their canopy I again found an interesting fresh leaf on the ground. But this leaf was neatly rolled up into a tube and carefully held together with silk. I picked it up, expecting to find a caterpillar inside when I unrolled it. And there it was indeed—a thin, pale microlepidopteran moth caterpillar. What surprised me was that the leaf roll was on the ground. Had it been discarded by the tree? Would a tree shed its leaves to get rid of any caterpillars rolled up in them?

I searched under the same tree and other poplar trees and within an hour or two had picked up 246 identically rolled-up leaves. Most of them had a similar caterpillar inside, about 0.3 to 0.4 inch long. Leaf-rolling by caterpillars is common, but finding rolled-up leaves on the ground under the tree is not. Every one of the rolled leaves that contained a caterpillar was missing most (though not all) of its stem—proof that the tree had *not* spontaneously aborted the leaves. Petioles are tough; they do not tear or break in a storm. The connection between the petiole and the twig would break first, and therefore it was not the tree that was getting rid of its caterpillars but vice versa.

The caterpillars had rolled themselves up by grabbing an edge of the leaf, attaching silk that stuck to the outer leaf edge, pulling in and attaching the other end of the (non-stretching) silk to the inner leaf surface, and repeating the process until they gradually rolled the leaf into a tube. After finishing that, they had reached through the top of the tube and gone to the considerable effort, and maybe risk, of sticking their neck out to chew through the petioles so that they and their leaf tube, or roll, would fall to the ground. They then stayed in their leaf rolls, and eventually pupated inside them. I kept each of the clipped-off leaf rolls with its enclosed caterpillar, which molted into a pupa, and in the first week of July little gray moths emerged.

Why did the leaf-rolling caterpillars clip off the leaves they were in? This leaf clipping was much different from what I had observed before. The other caterpillars stayed on the tree, where there were always other leaves to feed from. These, in contrast, were isolating themselves from the tree, and as a consequence seemed to be restricting their food supply. Why would that be advantageous? Are they safer on the ground than on the tree? To find out, I took 200 of the freshly dropped leaf rolls containing caterpillars, divided them into five groups, and distributed them to five different locations on the ground. One week later all the leaf rolls were still in place. So far, so good—the ground seemed safe. But maybe the caterpillars are equally safe on the tree. That would be difficult to know, though, if they normally don't stay there. I wondered, however, what would happen if I unrolled them from their apparently safe little houses after they had become grounded and then put them back up into the tree.

There were two young poplar trees in the clearing by my cabin, and I released numerous of the unrolled caterpillars onto their branches. They seemed not well suited to hanging on, especially to poplar leaves, which vibrate wildly in even the slightest breeze as though designed to shake off caterpillars, and many did immediately fall off. However, others hung on, and within a day I found thirty newly made leaf rolls. That is, my caterpillars almost immediately made themselves new homes. Two days later, however, seven of the thirty rolls had been clipped off. I saw no sign of predation, but it seemed that using tree saplings in a field was not a fair test to examine predation when their natural habitat is the crowns of large trees in the forest.

Making the observations was fun, especially if I could continue them in the top of a tree. So I carried a supply of leaf rolls that I had gathered on the ground up into the crown of an old aspen tree, perched comfortably on the branches, made sure that they had no existing leaf rolls or petioles from previously clipped-off leaves, and then unrolled one leaf at a time and released caterpillars onto these cleared, marked branches. Two days later I climbed back up and found thirty-seven new leaf rolls (presumably made by the more than eighty caterpillars I had released). One caterpillar was being eaten by a shield bug; eight partly rolled (or unrolled) leaves were empty; twelve partly rolled leaves each held a caterpillar, but none of the leaf had been eaten; seven fully rolled leaves with caterpillars inside had some of the leaf eaten; and there were nine clipped petioles. Thus, apparently several of the caterpillars had been eaten; being up in the tree was risky. These were interesting observations, but they could be interpreted in several ways, and no firm conclusion was possible—not enough for a scientific publication. Meanwhile, the caterpillars' season was over, and I thought about other things.

I did not get an opportunity to think about these caterpillars again until more than two decades later, when I started to write this book and found my notes squirreled away in a file. When I took a break and went for a daily jog during the next ten days in June 2006, I checked for freshly dropped leaf rolls on the gravel under the aspens along the road (in Hinesburg, Vermont). I found 208 of them. Twelve were freshly clipped and nine of these contained a mature larva. (As before, they pupated inside their rolls, and by the first of July the adults, the small gray moths who are fast runners as well as fliers, again emerged.) However, the remaining 196 leaf rolls had petioles (they had *not* been clipped off). All but two of these were without a caterpillar. Therefore, caterpillars had left them to make another roll, and they had done so apparently in time, before the tree had shed them.

The leaf rolls I had picked that had petioles but no caterpillars contained piles of caterpillar frass (feces), indicating that a caterpillar had been in residence for a long time, all the while feeding and fouling its nest (or pantry?). The leaf tissue on the inside of the rolls had turned color (yellow) or become necrotic. In short, these rolls had ceased to provide food, and they had been shed by the *tree*, presumably as a mechanism of getting rid of nonfunctioning leaves. However, before that happened, the caterpillars had left their deteriorating leaf rolls to seek fresh leaves and make a new roll to feed and hide in. That explained why many of the freshly clipped-off rolls that I had found earlier had little feeding damage inside yet contained a large caterpillar.

Apparently caterpillars leave their roll when it fills with feces or becomes necrotic, and then make another roll and resume feeding. As a result, many "empty" rolls accumulate on the trees and these fall off, but late in the summer. Finally, when the caterpillar is nearly mature, it clips off the *last* roll that it is in, and then rides with it to the ground, where it remains inside, pupates, and then emerges as an adult.

By late August I started seeing leaf rolls of another sort, on young basswood trees. Basswood leaves are gigantic relative to poplar leaves. To a diminutive (microlepidopteran) caterpillar that needs to roll up a leaf to hide in while feeding, a large leaf presents a problem. But these caterpillars had solved it beautifully. Each caterpillar had made a cut into the leaf all the way through several large leaf veins and then up to but not through another major vein. As a result a large portion of the leaf had flopped down while the rest of the leaf stayed up, and the dangling portion of the leaf was then rolled up. The leaf vein from which the roll then hung would continue to supply the nutrients to the rolled-up leaf used by the feeding caterpillar.

I continued to jog almost daily along the same road, enjoying the summer and at the same time keeping track of other signs of caterpillar magic. On 10 September I found out something new about caterpillars that I had never seen or heard about. There was at that time a caterpillar outbreak on the maples (both sugar and red). This outbreak was not as eye-catching as that of the fall webworms (who may enclose whole trees in diaphanous veils of webs) or that of the forest tent caterpillars. Still, it was eye-catching enough. On some maple branches as many as a third of the leaves were folded over, much as one might fold a piece of paper and then glue it together to make an envelope. It was the typical work of one of the thousands of species of microlepidopterans, and I was tempted to let it go at that—I knew a little caterpillar would be inside and feeding. So what? But for some obscure reason I idly opened several of these folded-over leaves, and to my great surprise saw neither caterpillars nor masses of feces. I was puzzled, because caterpillar feeding damage was evident—these caterpillars had eaten the underside of the leaves. Many birds learn to open leaf "envelopes," and these envelopes could have been easily opened. But if I had been a bird, I would have been disappointed—at least at first. I looked into many of them and wondered: where are the caterpillars?

No caterpillars were visible, but each leaf envelope contained a thin dark object an inch or more in length. At one end it was as thin as the lead in a mechanical pencil, and at the other end it was about one-tenth of

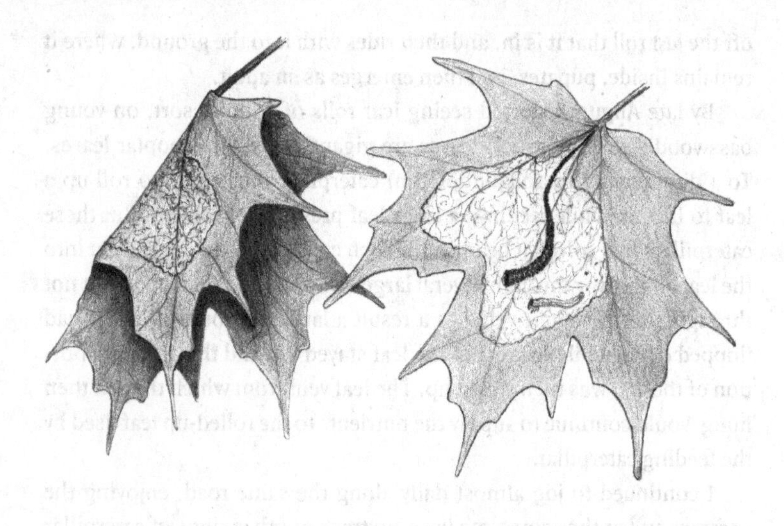

Fig. 18. Small moth caterpillar that feeds on the undersurface of leaves and makes a tube house in the space created by folding the leaf into a sandwich.

an inch wide and had an opening: it was a long, tapering tube. I eagerly broke several of these rather dry and friable tubes open, and again at first found nothing. However, I eventually did find what I was looking for: the tiny, almost transparent caterpillars. They were near the bottom of the tube rather than at the top. It turned out that they retreat to the bottom of their tube as soon as a predator opens the leaf envelopes they are in. But where did the tubes come from?

The tubes were made by the caterpillars, and from their own fecal pellets. Thus, these caterpillars were not soiling and spoiling their food, but instead using their own waste to make a retreat to hide in. They silked their fecal pellets onto the mouth of the "door" of their house, to gradually build a wider, longer tube as they grew. The beauty of this caterpillar's behavior is in the mind's eye, but in still another caterpillar it also enhances the fall foliage show.

It is mid-October. The aspen leaves have turned a rich deep golden yellow. On a slight breeze after a light frost, the leaves come tumbling down. Unlike those of the red maples, for example, where random blotches of yellow, red, purple, and pink may intermingle, the poplar leaves are uniformly, unvaryingly gold. But under some trees are exceptions—many of the bright yellow leaves have a conspicuous pea-green spot near the petiole, between the mid-vein and the nearby subsidiary vein. The color, and the very specific spot on the leaf where it is always located, catches the eye—you look around and find another, and another. And each time, the green spot is in exactly the same place on the leaf. The singular green spots in otherwise rapidly aging bright yellow leaves took me by surprise. They had to be caused by an outside agent. And they were.

Examining the green spots under a microscope, I could see through the transparent leaf epidermis, and beneath, within the leaf tissue itself, was a little pale green caterpillar with a trail of black fecal pellets behind it. This "leaf miner" caterpillar also rides the leaves down to the ground and then feeds on them. But it is too small to be eaten by birds, and it is probably also too small to be able to chew through a leaf petiole. This caterpillar could surely do its eating and growing early in the leaf's life cycle, but such a tiny caterpillar would very quickly dry out in summer heat and dry air. It would need to reach moist ground to pupate. It must leave the tree crown to do that, but for such a tiny larva to leave the leaf and enter the atmosphere in the dry summer heat could cause rapid death from desiccation. However, by shifting its caterpillar stage to late in the summer and early fall, when the weather is cooler and more moist and when the tree normally sheds its leaves, it can ensure itself a safe, moist harbor during a free ride to the ground. And just in case it still needs to feed a little more after it reaches the ground, as I assume it does, it persuades a part of the leaf to reduce its aging process, making it stay green. (See the last page of the color insert.) Color change signals leaf senescence and the beginning of fall. This caterpillar apparently has a chemical with which it manipulates the plant's genes to extend the life span and thus the length of the leaves' summer. Even though I never had a quarrel with fall, I think I could use a few drops of such a magic potion.

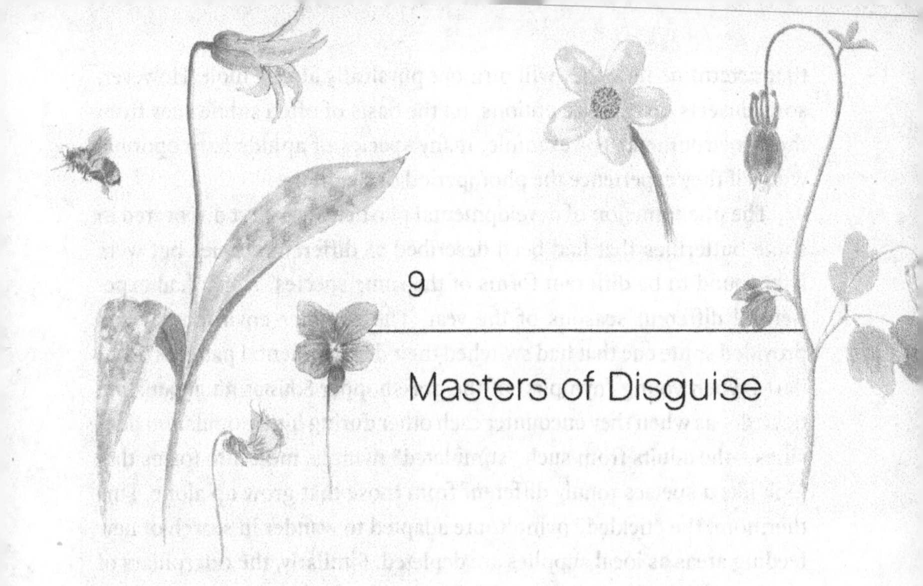

# 9

# Masters of Disguise

IT IS RARE FOR ANYONE NOT SPECIFICALLY LOOKING FOR a big plump caterpillar of a sphinx moth to find one. There is one exception, though—the tomato hornworms. We always have a tomato patch in our garden, and we almost always used to find several hornworms in it, although I have not seen any for years. The big green (and sometimes pale blue or black) horntail grub munching on tomato foliage in the garden transforms itself into a mummylike pupa encased in a hard brown shell and then remains in a deathlike stupor for the better part of a year in an underground crypt. The following summer it molts from its shell to be resurrected into a moth that flies only at night and feeds on the nectar of flowers, and superficially looks and acts like a hummingbird. But the moth is substantially more different from a hummingbird than a human is from an aardvark. Because they are unique and represent a strange breed that has been a favorite animal for unraveling many mysteries of development, I am always hoping to find one or several hornworms feeding on our tomato greens.

An insect's metamorphosis in body and behavior from larva to imago (adult) is amazing, yet it is easily taken for granted because of its inflexible inevitability.

It is difficult enough to envision butterflies exercising choice in behavior, much less to imagine their immature grubs exercising options that determine how they will turn out physically after a molt. However, some insects do exercise options, on the basis of often subtle cues from their environment. For example, many species of aphids have optional wings if they experience the photoperiod of a summer.

The phenomenon of developmental plasticity was first discovered in some butterflies that had been described as different species but were later found to be different forms of the same species, which had experienced different seasons of the year. The summer environment had provided some cue that had switched their developmental patterns. Similarly, when young (nymphs) of the grasshopper *Schistocerca gregaria* are tickled—as when they encounter each other during high population densities—the adults from such "stimulated" nymphs molt into forms that look like a species totally different from those that grow up alone. Furthermore, the "tickled" nymphs are adapted to wander in search of new feeding areas as local supplies are depleted. Similarly, the caterpillars of some species of moths also respond to their environment by switching one developmental pathway into another one, to produce forms that are adapted to reduce their chances of being eaten by predators.

The appearance of a caterpillar from one instar (a stage separated by a shedding of its "skin," or armor shell) to the next may commonly differ, but the new "uniform" that it wears is usually specific for all individuals. However, in some species there are two or more options for the "uniform," depending on what the caterpillar experienced when it was younger. For example, in one sphinx moth species, *Laothoe populi*, when the caterpillars are raised on a white background they molt from green to white. In another moth, *Nemaria arizonaria*, when the young caterpillars perch and feed on oak catkins in the spring they resemble what they eat. Later—by summer, when they perch on twigs and eat leaves—they molt into a new form that resembles twigs.

For sheer number and variety of disguises in the same species, I vote for the Abbott's sphinx moth, *Sphecodina abbotti*. Its caterpillar transforms inself through an amazing series of four disguises—a noxious insect, two different kinds of camouflage, and a snake. I made the acquaintance of these caterpillars at the University of Minnesota biological station at Lake Itasca, where I found three different forms of the caterpillar on the same food plant, wild grape.

The first four instars of all *abbotti* caterpillars are chalky white and thus conspicuous on green grape leaves. However, they curl up to resemble larvae of a cimbicid wasp. Cimbicid larvae are chemically protected—they can squirt a defensive fluid from glands along their sides. Apparently the young *abbotti* caterpillars mimic this distasteful model, because they do not have a "horn" at the end of the abdomen like other sphinx moth caterpillars (hence the common name, hornworms); instead, that structure is altered to look like a yellow translucent drop of fluid. It is unlikely that color, structure, and behavior would all converge to accidentally mimic the wasp larva, especially since the caterpillar's appearance changes not partially but drastically when it molts into its last larval instar.

More startlingly, rather than just changing into a single totally different "uniform," it morphs into either of two possible forms that are not only different from the previous form but also strikingly different from each other. One form is brown and streaked with black. This form is nearly invisible on a background of brown grape bark, where it hides in the daytime. At night it comes up to the leaves, and after finishing feeding on a grape leaf it snips off the remainder of the uneaten leaf and crawls back down and hides, staying immobile throughout the day while pressing itself tightly against the old growth of flaking grapevine bark. The other, rarer form of the same (fifth) instar of the same caterpillar on the same plant has large luminous pea green patches on its back and along its sides. This form feeds in the daytime and does not perch on old grapevines that have bark; instead, it stays on the young, smooth green grape stems.

The adaptive significance of simultaneously having two wildly contrasting forms of the *abbotti* sphinx moth in the last instar caterpillars on the same food plant is obscure. The brown form is clearly, in both appearance and behavior, adapted to hide on bark of grapevines (and Virginia creeper, another of its food plants). But the visually striking form with the green patches appears to be an anomaly with an as yet unknown selective advantage. I speculate that it is so different from the other that a predator who finds one form may be too distracted to search for and see the other. As already mentioned, a bird that has found a particularly juicy morsel will search for others that look like it. If a bird finds one caterpillar form on a grape plant—say, the bark mimic on a grape stem—it will search for others of the same type and in the same setting. It will thus, by

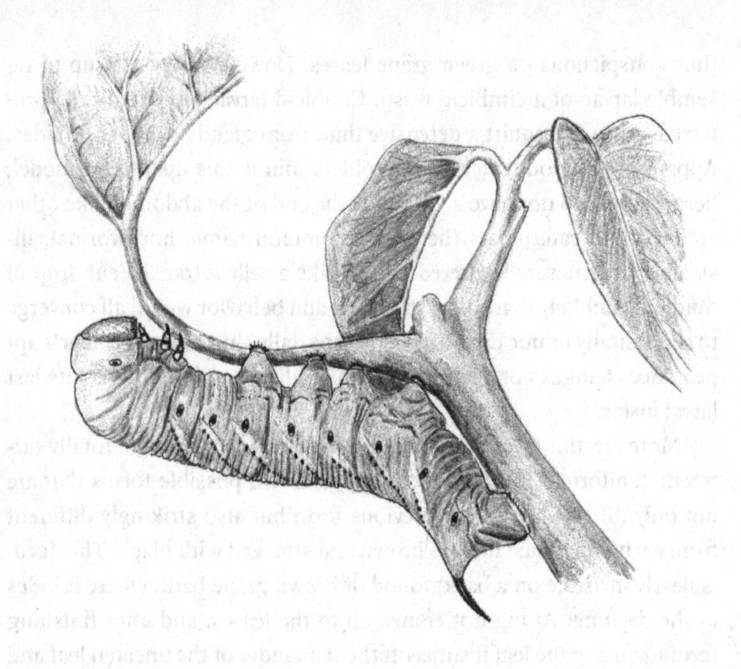

Fig. 19. The hind end of most sphinx moth caterpillars has a "horn," as in this tobacco hornworm, *Manduca sexta*.

knowing what it is looking for, more easily miss what is different. This is indeed the effect I tested with the students and the poplar sapling: some students searched an hour before finding the first caterpillar, but then after they found the first one they almost immediately found the second. A very common caterpillar, no matter how well camouflaged, is likely to be found eventually, and it is thus dangerous for an edible caterpillar to be on a bush with other edible company of the same appearance. However, one that is wildly different from that company has a good chance of not being noticed.

Neat as such tricks of the caterpillars in their game of hide-and-seek with predators and parasitoids might be, the question of mechanism always arises. How can two different morphs of the same age manage to be on the same food plant at the same time? Do some moths lay eggs that develop into only one form, whereas other individuals of the same moth species lay eggs that develop into the other form and then randomly lay

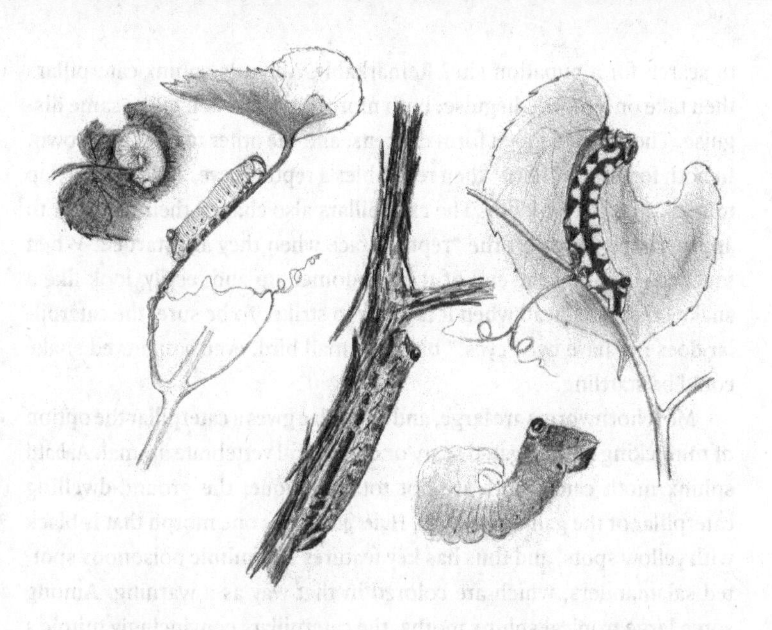

Fig. 20. **Four disguises of Abbott's sphinx caterpillar. The typical hornworm horn is adapted to look like a drop of yellow fluid in the young caterpillar; later, it looks like an eye in the mature larva, which mimics a snake.**

their eggs on the same plant? Or are there instead two different morphs contained within any one individual moth? Alternatively, the different morphs could result from a developmental switch that is activated by an external cue, perhaps the mere presence of other caterpillars in the vicinity. A caterpillar cannot readily afford to leave its food plant, but changing its disguise may be almost equivalent, or even better, because it still retains the food. The presence of others could be the cue for such induction into the other (rarer) morph, since crowding causes a dramatic color change in another sphinx moth caterpillar, *Erinnyis ello* (Schneider 1973).

Both of the final instar forms of the *abbotti* sphinx caterpillar rely on blending in so as not to be noticed, and this requires that they don't move. No matter how well camouflaged a caterpillar is, it is likely to become dead meat if it moves when a bird is nearby. But what happens when the larva leaves the food plant and must crawl along the ground in order to search for a pupation site? Remarkably, Abbott's sphinx caterpillars then take on a fourth disguise: both morphs now switch to the same disguise. The mottled green form darkens, and the other form stays brown. In both forms the "horn" then resembles a reptile's eye, and the anal flap mimics a reptile's mouth. The caterpillars also change their behavior to appropriately show off the "reptile" face when they are startled. When touched, they curl the end of their abdomen up and, eerily, look like a snake raising its head when it is ready to strike. To be sure, the caterpillar does not have two "eyes," but to a small bird, even a one-eyed snake could be startling.

Most hornworms are large, and large size gives a caterpillar the option of mimicking an elongated scary or distasteful vertebrate animal. *Abbotti* sphinx moth caterpillars are not totally unique; the ground-dwelling caterpillar of the gallium sphinx, *Hyles gallii*, has one morph that is black with yellow spots, and thus has key features that mimic poisonous spotted salamanders, which are colored in that way as a warning. Among some large tropical sphinx moths, the caterpillars convincingly mimic a snake's head (Miller et al. 2006), but in this case the effect of scales on the head comes from the folding of the front legs on the caterpillar's underside, which is turned up in its snake display. The "eyes" (in this case, two of them) are derived from puffing out darkly pigmented skin on the sides of the head end. That is, the head of this snake mimic is fashioned from the front end, rather than the back as in *abbotti*. The snake head display is, in a butterfly, even found in the next stage, the chrysalis (Aiello and Silberglied 1978).

Coloration is an important component of many disguises, but color as such can also have another, equally important function. Dark coloration serves both to protect an animal from the sun and to increase the absorption of solar energy. Many butterflies are colored in ways that facilitate solar heating, which permits their flight muscles to operate in a cold environment. In caterpillars, instead, elevated body temperature speeds up the growth rate and greatly shortens the time until the relatively safe pupal stage is reached. In caterpillars, growth rate is perhaps one of the most critical factors in avoiding predators, because every day that the caterpillar stage can be shortened is a day when the gauntlet of both parasites and predators is avoided. Being black and exposed to the direct sun is, however, a two-edged sword. At the same time that it

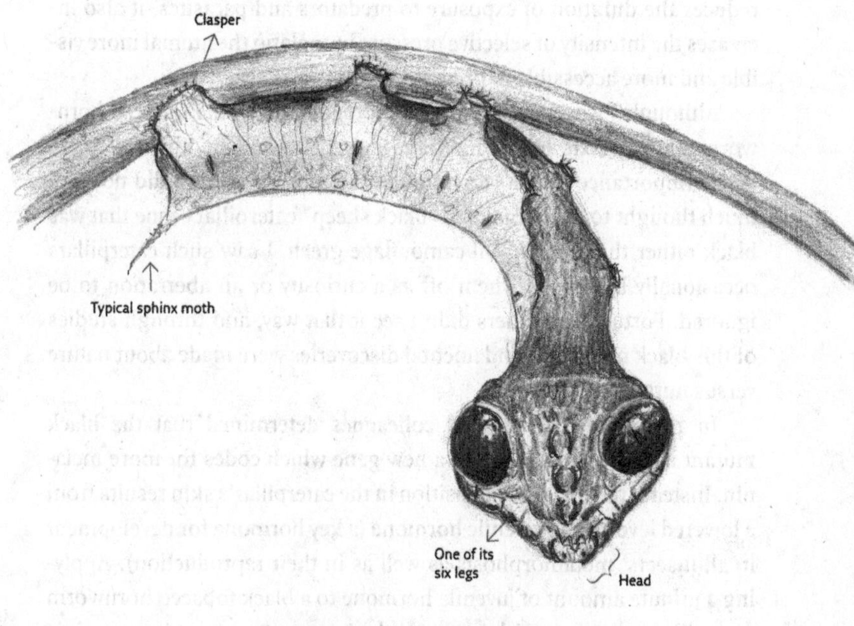


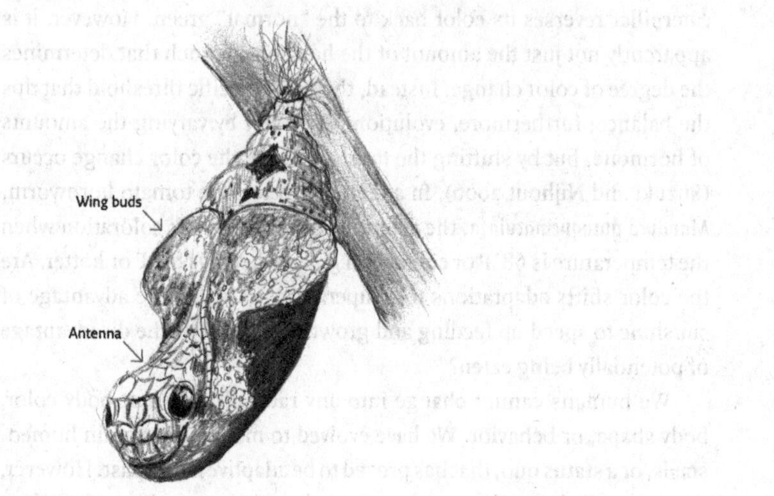


Fig. 21a and b. **Snake head display of the caterpillars of the Central and South American sphinx moth, *Hemeroplanes triptolemus*; and of the chrysalis of the butterfly *Dynastes darius* (drawn from photographs in Miller et al. 2006).**

reduces the duration of exposure to predators and parasites, it also increases the intensity of selective pressure by making the animal more visible and more accessible to predation.

Although I was working on thermoregulation of the tobacco hornworm *Manduca sexta* sphinx moths as a graduate student, and was aware of the importance of color in butterfly thermoregulation, I did not give much thought to the occasional "black sheep" caterpillar—one that was black rather than the typical camouflage green. I saw such caterpillars occasionally but passed them off as a curiosity or an aberration to be ignored. Fortunately, others didn't see it that way, and through studies of this black mutation fundamental discoveries were made about nature versus nurture.

In 1973, Jim Truman and colleagues determined that the black mutant is not just the result of a new gene which codes for more melanin. Instead, the melanin deposition in the caterpillar's skin results from a lowered level of the juvenile hormone (a key hormone for development in all insects' metamorphosis as well as in their reproduction). Applying a minute amount of juvenile hormone to a black tobacco hornworm caterpillar reverses its color back to the "normal" green. However, it is apparently not just the amount of the hormone as such that determines the degree of color change. Instead, there is a specific threshold that tips the balance; furthermore, evolution works not by varying the amounts of hormone, but by shifting the *threshold* where the color change occurs (Suzuki and Nijhout 2006). In a related species, the tomato hornworm, *Manduca quinquemaculata*, the caterpillars develop black coloration when the temperature is 68°F or colder and green when it is 82°F or hotter. Are the color shifts adaptations to temperature, in which the advantage of sunshine to speed up feeding and growth rate exceeds the disadvantage of potentially being eaten?

We humans cannot change into any radically different body color, body shape, or behavior. We have evolved to maintain a certain homeostasis, or a status quo, that has proved to be adaptive in the past. However, the genes of a butterfly are the same as those in a caterpillar. The difference is which are turned on or off, and when. It's all "environment" as well—in this case mostly the internal environment that keeps changing through development. Whenever I see a weight lifter, a runner, a mathematician, an actor, a sumo wrestler, or a dancer, I am reminded that, like caterpillars, we do have the capacity to accomplish amazing changes, and these are sometimes in response to simple, subtle cues, which are like switches that control development. No one develops with a complete predetermination of the features and faculties that he or she will eventually come to "own." To the contrary, although we are built from pretty much the same blueprint, many of our specific, individual "talents" can be activated only if we exceed a certain threshold of effort, but probably this threshold is also specific to each individual. I was reminded of this while training to metamorphose from a slow cold-weather-adapted animal that conserved energy and heat to one who could expend energy at high rates and dissipate heat as fast as possible. If in a caterpillar a mere visual stimulus can change gene expression affecting development, then why not exercise in us?

Once an end result is achieved, it is hard for us to imagine an alternative that has proceeded along a different developmental trajectory without crediting it to magic or "talent." When we see in others something that we find incomprehensible for ourselves, it is easy to pass this off as "genetic." Naturally, it is exactly that; but this description still omits the essence of development, the miracle on the miracle. The possibility of individual caterpillars to generate amazingly different forms makes me appreciate what is possible in the debate over nature versus nurture. Much of what we are and become depends on minute subtleties, and that gives me hope in the reality of free will, and its power if we choose to exert it.

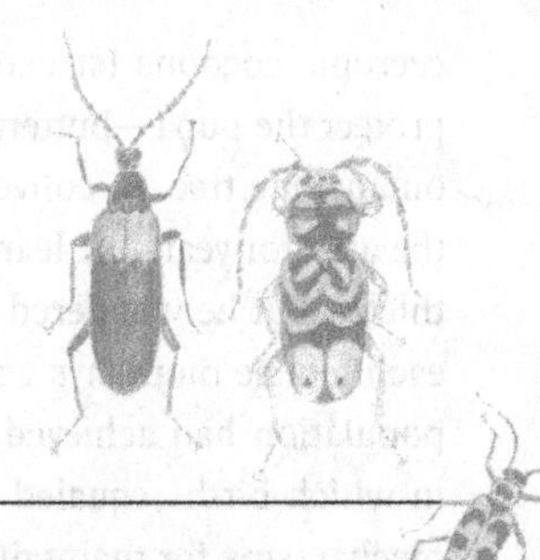

# 10

# Cecropia Moths

22 June 2007. THE TREES ARE FINALLY FULLY LEAFED OUT and have completed most of their twig growth now, and—more noteworthy—the buds that had produced the twigs with new leaves have in some species also already produced the buds for next year's growth. And a few of these new red oak buds have already "broken" to produce a second spurt of twig and new leaf growth a year ahead of the others on the same branch. There is now ample foliage, and the giant silk moths have laid their eggs. I pick up a dead luna moth in the woods. Its pure white body "fur" contrasts with its fresh leaf look from its large, pale greenish blue wings. One of its two "tails" has broken off and the edges of its delicate wings are frayed from its hectic flight during the last week's nights—the only time allotted to it as an adult. This is the only narrow window of time that one can meet these gorgeous creations as adults. This moth's abdomen is shrunken—it had managed to lay its eggs, and its green caterpillars are probably hatching and starting to feed on the new oak, maple, and birch leaves.

IN HIS MASTER'S THESIS, A STUDENT, FRANK L. MARSH, wrote: "About the middle of March, 1933, the writer chanced upon an area within the limits of southwest Chicago" in which he discovered the cecropia cocoons (structures of silk made by caterpillars to hold and protect the pupa—butterflies don't have cocoons) visible by the dozens on any one tree. In conversations with several people who had lived in the area for years, he learned that the cocoons had "always been just as thick." But he wondered why they were not even more common, since each female moth lays 200 to 400 eggs. Marsh surmised that the moth population had achieved and was maintaining a state of equilibrium, in which births equaled deaths. He then proceeded to study possible mechanisms for maintaining such an equilibrium. He concentrated on the causes of death that could be deduced from the contents of the 2,741 cocoons that he collected. His was a project that I can scarcely conceive of, since I feel lucky to find even one of these now very rare cocoons; I have seen perhaps three in the last five years.

With all the abundance of the northern forest caterpillars, it is easy to forget that most of them turn into moths (primarily of the families Noctuidae and Geometridae). Not only are the moths much rarer than their larvae—there is perhaps only one moth per 100 larvae—they are almost all nocturnal. Summer nights belong to the moths and fireflies. The difference is that we see the fireflies. I "see" the moths only in my mind's eye—especially the big ones of the family Saturniidae, the giant silk moths, which can be easily mistaken for bats when they fly in the dark. We know they are out there mainly because we may find their caterpillars in the summer, and for several species, such as the cecropia moths and the prometheans, also their cocoons in the winter.

The woods of New England contain (or contained) an abundance of half a dozen species of the spectacular saturniid moths. All of them are decorated in striking color patterns and clothed in a fine, though thick, velvety "fur" (modified scales, technically pile) that not only gives them their bright and intricate color patterns but also insulates them after they shiver to warm up to get ready for flight.

The cecropia moth, *Hyalophora cecropia*, is the largest of the local wild silk moths, and its cocoon is a brown baggy spindle-shaped structure. Although these moths' caterpillars have been tried for commercial silk production, they have been used much more successfully as laboratory animals in numerous studies that have revealed secrets of the physiological connections of neurons and hormones affecting behavior, development, and metamorphosis. The Harvard biologists Carroll Williams, Jim

Truman, and Lynne Riddiford are legendary and strike me as scientific shamans because of their extraordinary clever and revealing experiments, which delved deeply into the mysteries of the reincarnation of a caterpillar into a moth, or presumably any insect's metamorphosis from larva to adult. Among their numerous discoveries was that behavior patterns are inscribed on neurons and are expressed under the influence of hormones. Their studies also showed that internal and external (environmental) stimuli, filtered through the central nervous system, affect the body profoundly. Our vertebrate line split away from that of the insects at an early stage of evolution, but we still share many basic mechanisms, including those discovered in moths. These mechanisms differ not so much in kind as in degree and in how and where they are applied.

Cecropias emerge from their pupae as adults when they shed their pupal "skin" (strictly, an exoskeleton), and then they crawl out the escape hatch of their cocoons around noon on any one day in May. The freshly emerged moths then hang stationary, expand their soft flaccid wing stubs (the outlines are visibly inscribed on the hard pupal exoskeleton), and inflate them with blood to stretch them until they have expanded to full size. The still soft fresh moth then releases a hormone from the brain that activates a hardening agent, and the wings then become frozen into their final shape. When eclosion—the act of emerging from the pupa, which is triggered by hormones—is finished, the moth then purges its gut. What is purged, the meconium, contains fecal and urinary wastes that had accumulated during the pupal stage (which lasts more than ten months); in the females, the meconium also contains the sexual attractant.

Males, who search for females solely by their scent, may come from miles away, flying upwind. Mating in this species begins just before sunrise and lasts about fifteen hours. (Most of the "mating" time is actually guarding, in which the male prevents another male from mating with his female.) Egg laying follows immediately, and the female then flies every night for about a week to scatter her eggs in clusters. They are coated with a glue and when deposited stick to the leaf undersides of several species of forest trees.

The larvae hatch after about twelve days, near 1 June, and then they grow through five stages. Each of these "instars" is separated by a molt. The first larval instar is black and is covered with orange and black tubercles. The second has bright orange and black spots. The third is yellowish green with black spots and blue tubercles. The fourth is light green with a broad dorsal band of blue and yellow, coral and black tubercles. The fifth larval instar is also light green but with a dorsal band of only blue. After each molt the larva eats its old, shed skin, except for the spines and tubercles.

By mid-July the mature larvae stop feeding and initiate violent gut peristalsis that empties them. They are then restless and wander, often leaving the food plant. They eventually stop wandering and begin an incessant labor of about a week, by day and night, to spin the cocoon. First a larva makes the outer shell, leaving an exit valve for the moth's eventual escape. It then makes the inner cocoon layer by rapid zigzagging sweeps of the head—all except for the threads at the exit valve, where it lays down the silk in parallel lines with the long axis of the cocoon. The larva keeps turning from one end of the cocoon to the other, first laying down silk and then saturating the entire structure with saliva that cements the threads together and makes the cocoon tough and waterproof. The cocoons of this species are unique in having two layers, an outer and an inner section, and an escape hatch at one end for the moth to emerge from during the following summer. The double case probably helps to protect the defenseless pupa from predators; a would-be predator has to invest considerable effort to penetrate even the first, outer wall, and if it manages to do so and then finds only another wall, it may leave.

After working ceaselessly for several days to finish the cocoon, the caterpillar orients itself within this cocoon so that its head faces the exit valve. It then becomes quiescent and begins to shrink as it reorganizes its body and finally sheds its last larval skin to become the pupa. In the fall and winter after the leaves are down, the cecropia cocoons are conspicuously revealed because they are attached to bare tree branches. The pupae hibernate and can survive in a frozen state, and the moths emerge the following summer provided that the brain has experienced a prolonged period of cold; cold is necessary to activate the brain, telling it that winter has occurred. As a consequence, this moth can produce only one generation per year—unlike some other species that don't need chilling and that can have at least two broods in the south, where summers are longer.

This scenario of development from egg to moth is the normal one;

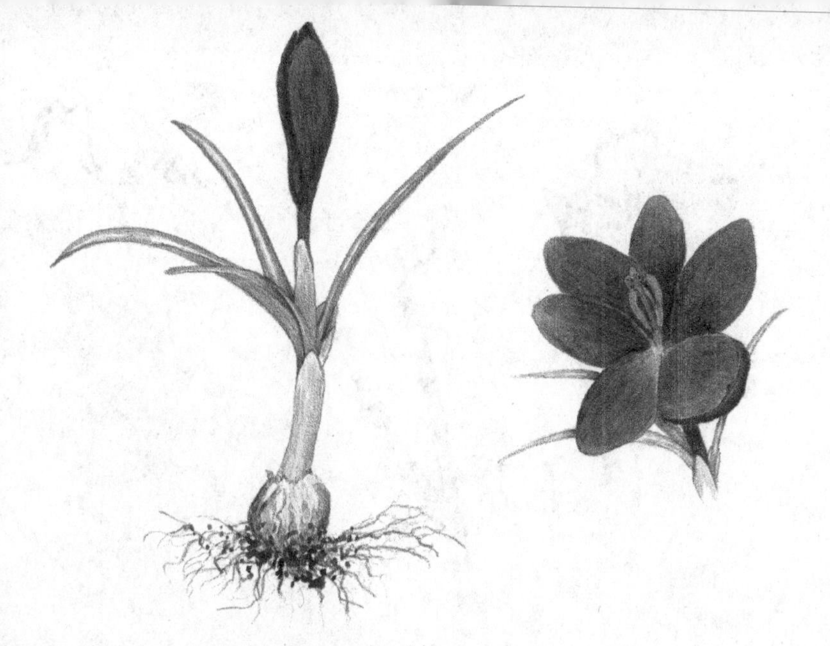

1a. Crocus flower during the daytime, as opposed to night.

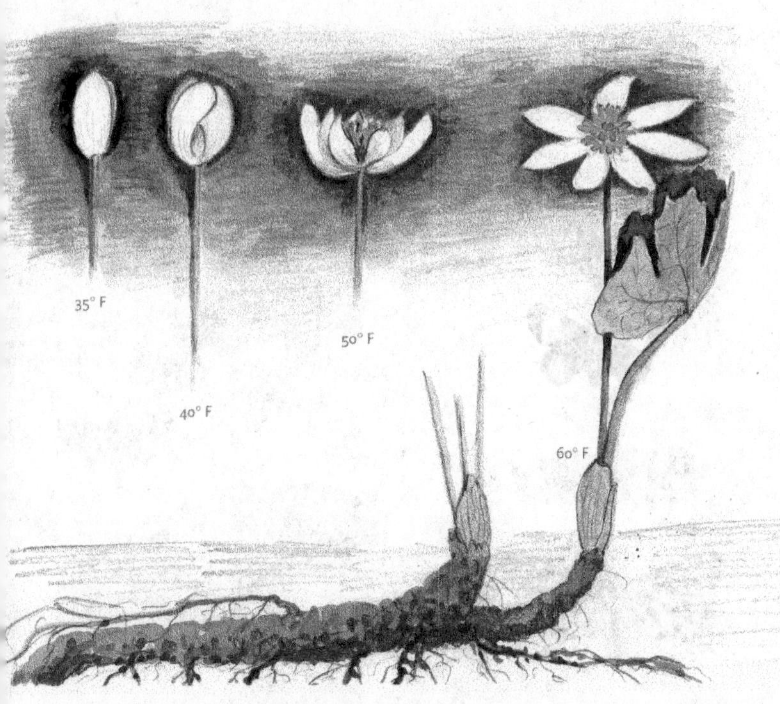


1b. Bloodroot flower as a result of temperature.

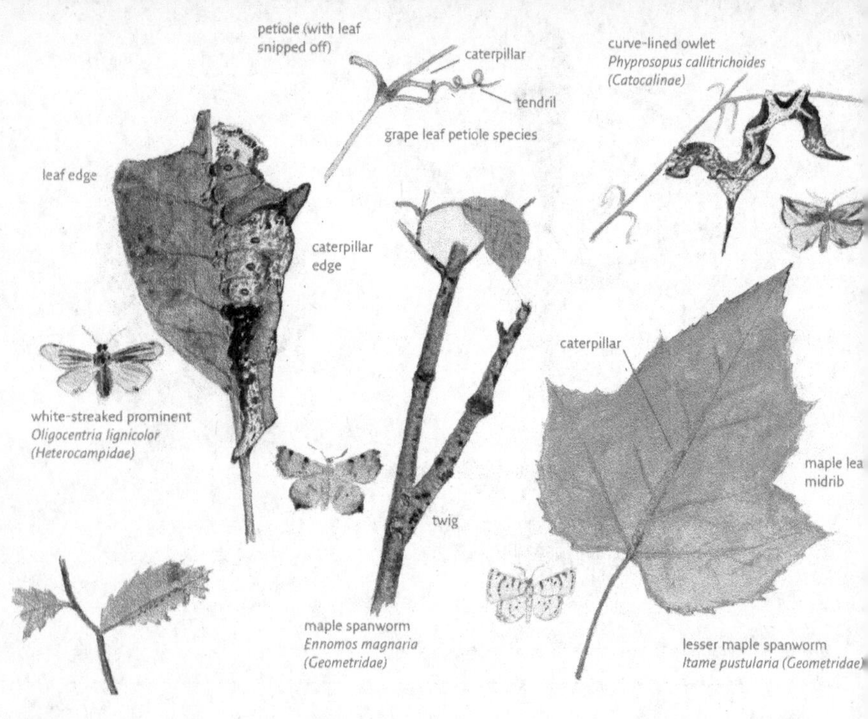


Fig. 2a. **Caterpillars eaten by birds. Try to find the six different caterpillar disguises: two leaf-edge mimics, one petiole mimic, one twig mimic, one leaf midrib mimic, and one possible debris mimic. Four of the corresponding adult moths also palatable to birds have been included.**

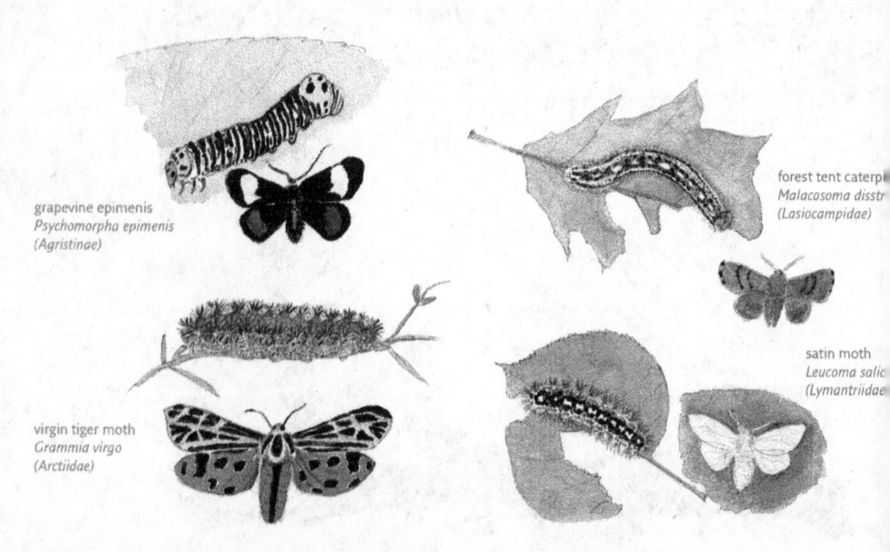


Fig. 2b. **Four different "unpalatable" caterpillars and their adults (imagos).**

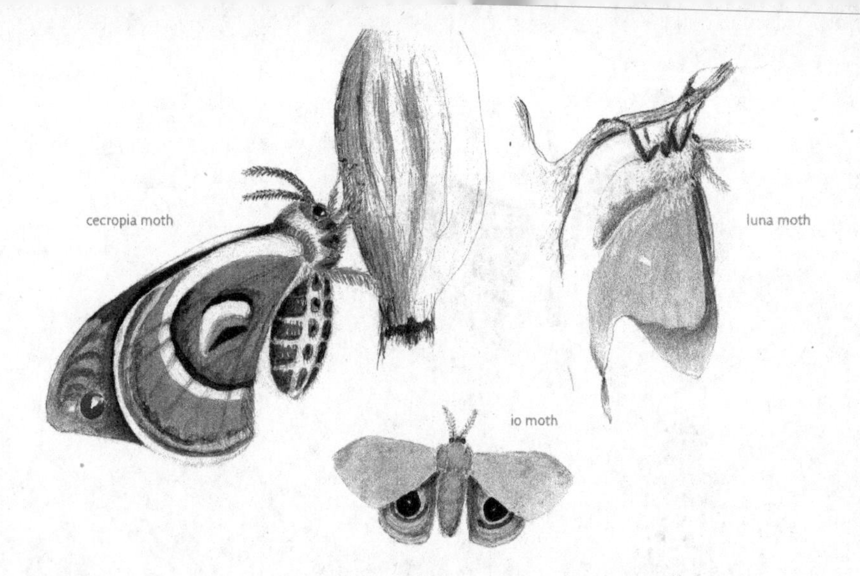


Fig. 3a. A cecropia moth on a cocoon from which it has just emerged. The cecropia is one of the better-known saturniid silk moths, to which the luna *(right)*, the io *(lower center)*, polyphemus, and promethean also belong.

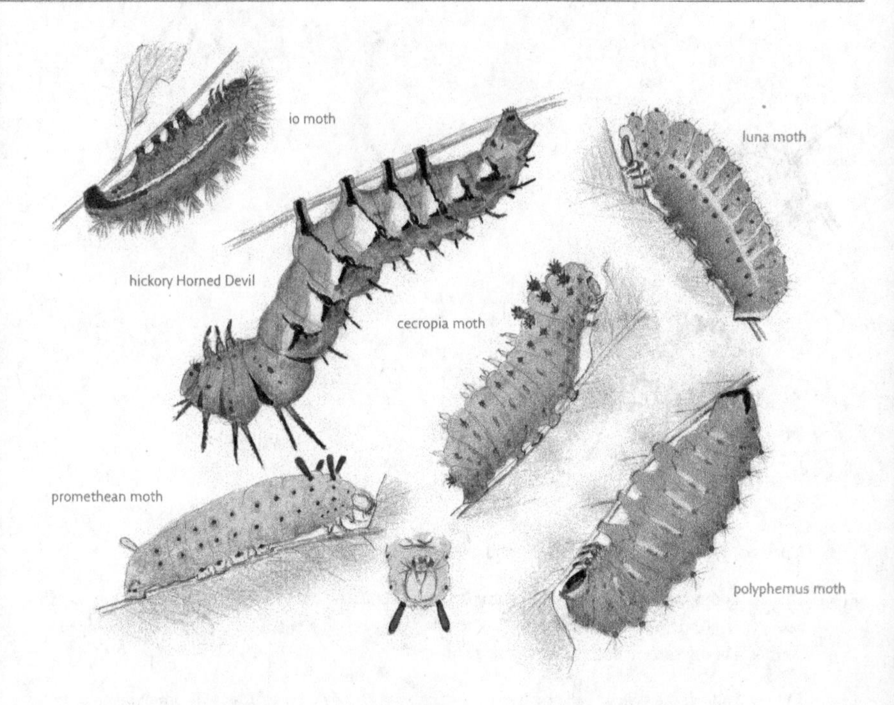


Fig. 3b. Caterpillars of North American saturniid moths.

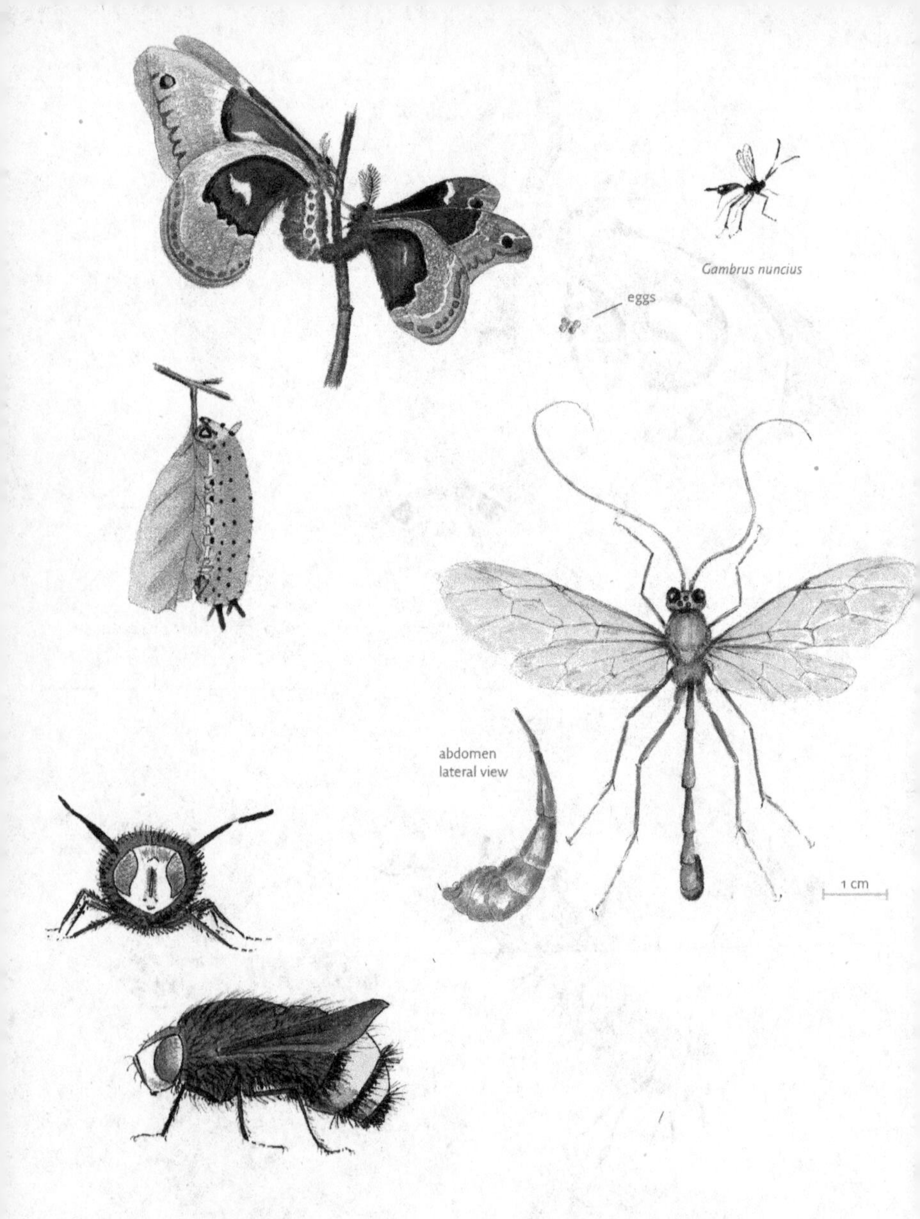


Fig. 4a. A pair of promethea moths *(above, top)* and one of their caterpillars.

Fig. 4b. The ichneumon wasps, *Enicospilus americanus* *(above, middle)*, and *Gambrus nuncius* *(upper right)*, which parasitizes promethean moths. Only one of the first develops in a caterpillar, but usually twenty to forty (or more) of the second emerge from a single moth.

Fig. 4c. The big white-faced yellow-bellied tachinid, *Belvosia bifasciata (lower left)*, a silk moth parasite that showed up near my camp for the first time in 2006.

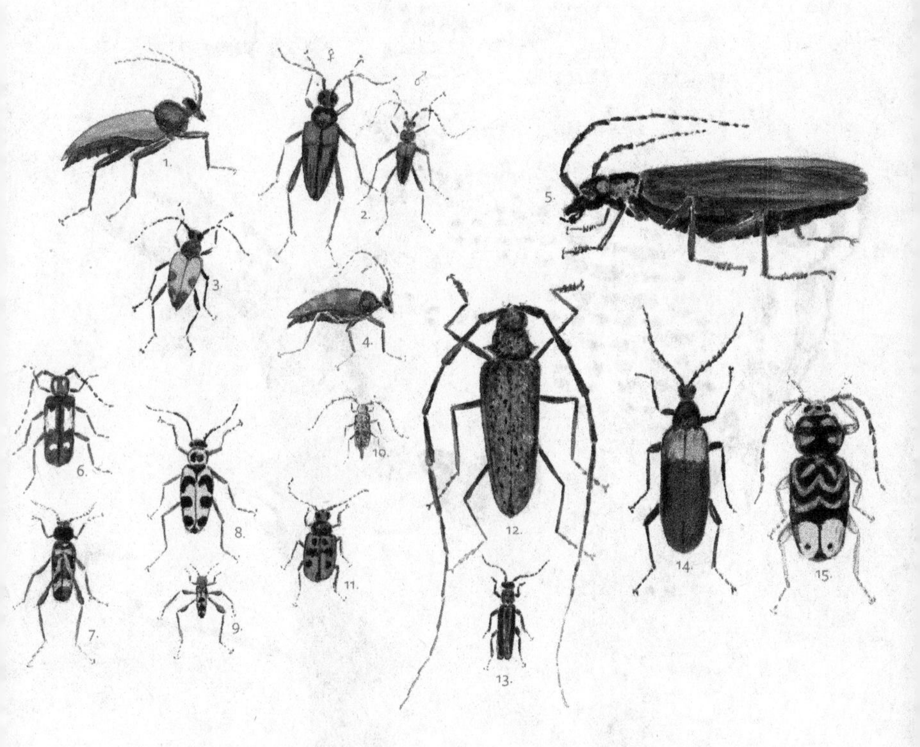


Fig. 5. A sampling of longhorn beetles from northeastern North America.
**1.** *Cosmosalia chrysoma.* **2.** *Stictoleptura canadensis.* **3.** *Judolia quadrata.* **4.** *Typocerus confluens.* **5.** *Stenodontes dasytomus* (spotted stem borer). **6.** *Saperda cretata* (spotted apple borer). **7.** *Clytus ruricola.* **8.** *Strophiona nitens* (chestnut bark borer). **9.** *Neoclytes approximates* (redheaded ash borer). **10.** *Urugraphis despectus.* **11.** *Tetraopes tetrophanus* (red milkweed borer). **12.** *Monochamus notatus* (northeastern sawyer), male. **13.** *Oberea affinis.* **14.** *Desmocerus palliates* (elderberry borer). **15.** *Glycobius speciosus* (sugar maple borer).

Fig. 6. Life at a yellow-bellied sapsucker's sap lick. Insects and hummingbirds also are attracted to the birch's sap.

g. 7a. The plants that I picked and sketched on 12 May 2007. The leaves of seven of them are those retained from the previous summer.

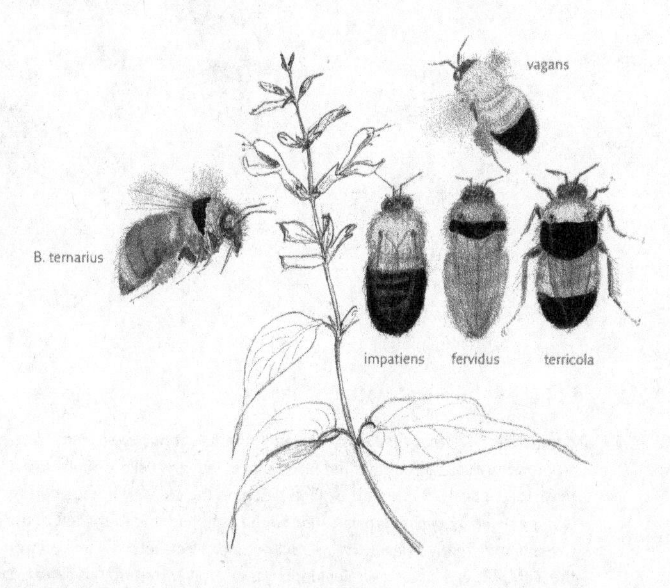


g. 7b. The common bumblebee (*Bombus*) species that I expected to see.

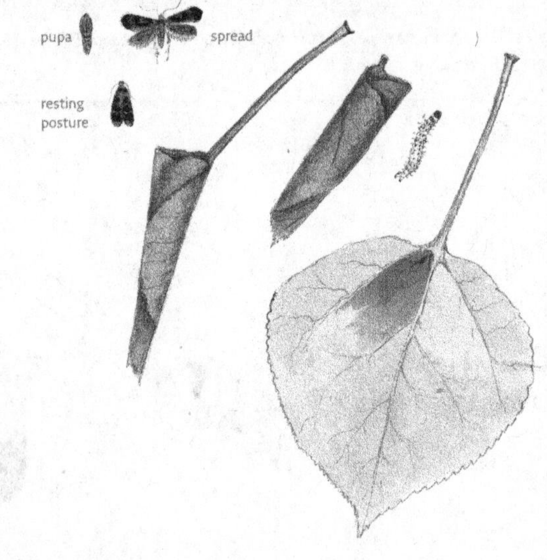


Fig. 8a and 8b. A bunchberry (*Cornus canadensis*) flower in the fall, about five months after the usual time of bloom for this species. Bright red fallen maple leaf and yellow poplar leaf are to the top left and right, respectively. But note the green patch on the old weathered poplar leaf below the flower. It was caused by two caterpillars, one to the right and one to the left of the midrib, feeding insi the leaf and greatly delaying its senescence (see also sketch, where only one caterpillar is inside the leaf). A much larger microlepidopteran caterpillar rolls up the poplar leaves and feeds inside the rolls instead (see chapter 8, "Artful Diners").

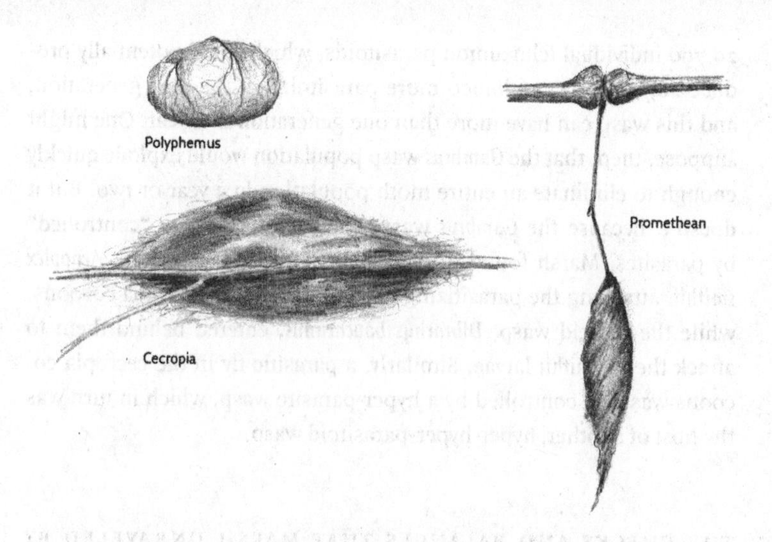


Fig. 22. Cocoons of cecropia, polyphemus, and promethean moths. The first has a double wall, is attached to twigs, and has an exit sleeve at one end. The second has a single wall, is built inside a rolled leaf, and has no exit hole—the moth exits by dissolving a hole with enzymes in its saliva. The third is also rolled in a leaf, but it does have an exit hole and it is attached to a twig by a long silk strap (shown is a cocoon that is at least a year old; the silk ring around the twig has constricted its growth).

however, it is actualized in only a small fraction of the caterpillars that hatch, perhaps about one in fifty to 100. Few survive to the pupal stage, and fewer still to the adult stage. We have little idea about the mortality of the eggs and caterpillars, but the pupae can be collected, and they show whether or not a moth emerged and what happened in those cases where no moth emerged. Of the 2,741 cocoons that Marsh collected and analyzed, 10 percent were chewed and pecked open and the contents had been removed; they were the victims of deer mice and woodpeckers. The other 90 percent of the mortality was caused by parasitic flies and wasps. Three percent of the pupae had been destroyed by flies (of the family Tachnidae). Twenty-three percent were victims of a species of ichneumon wasp, *Spilocryptus extrematus* (the genus is now renamed *Gambrus*).

Each of the 630 pupae killed by *Gambrus* hatched not just one wasp but, on average, thirty-three. That is, these pupae represent 630 × 33 =

20,790 individual ichneumon parasitoids, which could potentially produce 20,790 × 33 = 686,000 more parasitoids in the next generation, and this wasp can have more than one generation in a year. One might suppose, then, that the *Gambrus* wasp population would explode quickly enough to eliminate an entire moth population in a year or two. But it doesn't, because the *Gambrus* wasps are themselves also "controlled" by parasites. Marsh found, for example, another ichneumon, *Aenoplex smithii*, attacking the parasitizing *Gambrus* inside the cecropia cocoons, while the chalcid wasp, *Dibrachys boucheanus*, entered behind them to attack the *A. smithii* larvae. Similarly, a parasitic fly in the cecropia cocoons was also controlled by a hyper-parasite wasp, which in turn was the host of another, hyper-hyper-parasitoid wasp.

THE CHECKS AND BALANCES THAT MARSH UNRAVELED BY patient rearing of pupae collected in the field give us a glimpse of only some of the links that make up an ecosystem, which extends from the microscopic to the top predators. He also looked further into the mechanisms of the parasite-host relationships that reveal the subtlety of the tactics in the arms races between parasites and their hosts.

Timing is important. *Gambrus* wasps, for example, are attracted to their hosts, the caterpillars, by the odor of fresh silk these caterpillars exude while spinning their cocoon. The wasps arrive as soon as cocoon spinning starts, and to have a chance to lay their eggs they must be there while the cocoon is still soft. Otherwise they cannot thrust their ovipositor in to insert their eggs. Nevertheless, Marsh counted more than 1,000 *Gambrus* eggs in one recently spun cecropia cocoon, whereas on average only thirty-three larvae could grow to adulthood in one caterpillar. He therefore concluded that when there are too many larvae, the excess is reduced by cannibalism.

Marsh was correct, but despite the great complexity he discovered he still greatly oversimplified the "real" world of the moth's population dynamics. At that time he could not have considered an even more fantastic phenomenon of some ichneumon parasitoids: polyembryony, in which one egg can divide to produce multiple genetically identical individuals. That is, the egg clones itself to make as many as 100 or more individuals, enough to consume the whole caterpillar. If two wasps each lay an egg into the same caterpillar, then there will be two clones—each with many individuals—developing simultaneously. And, in a newly discovered twist, some of the larvae of any one clone may be precocious and soon die, but not until they have acted as mobile jaws to kill possible competitors of other clones of both the same species and different species.

The resulting checks and balances of predators, parasites, hyperparasites and hyper-hyper-parasites, cannibalism, and disease organisms ensures that the summer mortality depends on population density (it is density-dependent), so that no one population can completely eliminate the other, and the forest—the other end of the chain of cause and effect—stays green all summer long.

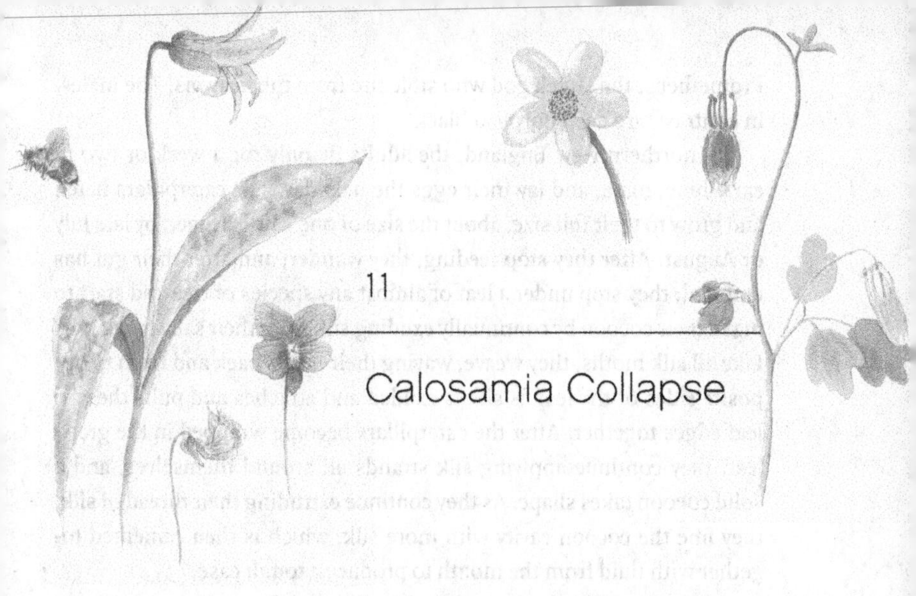

# 11

# Calosamia Collapse

IN 1903, W. J. HOLLAND, THE MOTH MAVEN OF NORTH America, wrote this about the promethean moth, *Calosamia promethea*, in his "bible" of the American moths: "Every country boy who lives in the Atlantic states is familiar with the cocoons, which in winter he has found hanging from the twigs of the spice-bush, the sassafras, and other trees."

I am lucky indeed to have made the aquaintance of these giant silk moths and their cocoons. But country living is apparently not what it used to be—I have not met any country boy, and only the very rare biology graduate student, who has even an inkling of what a promethean moth is. Most students don't know a cocoon from a pupa. Perhaps that's not surprising, because they haven't seen any, either in their everyday lives or in school.

Prometheans were still in abundance recently, near my camp in western Maine, where almost every summer their handsome blue-green caterpillars fed on the leaves of ash and cherry trees. Like many other saturniid silk moth caterpillars, these are strikingly beautiful when viewed from up close. They are decorated with four bright crimson tubercles on their front end and have splashes of yellow along their pale bluish green sides. The adult moths are beautiful, too. The females have fiery orange-red wings with white margins—hence the name promethean, for

Prometheus, the Greek god who stole fire from the heavens. The males, in contrast, are deep purplish black.

In northern New England, the adults fly only for a week or two in early June, mate, and lay their eggs the next day. The caterpillars hatch and grow to their full size, about the size of one's little finger, by late July or August. After they stop feeding, they wander; and after their gut has emptied, they stop under a leaf of almost any species of tree and start to make their cocoon by continually exuding silk from their salivary glands. Like all silk moths, they weave, waving their heads back and forth to opposite sides of the leaf as silk is exuded and attaches and pulls the two leaf edges together. After the caterpillars become wrapped in the green leaf, they continue applying silk strands all around themselves, and a solid cocoon takes shape. As they continue extruding their thread of silk, they line the cocoon cavity with more silk, which is then cemented together with fluid from the mouth to produce a tough case.

Caterpillar silk is both tough and flexible, and for centuries it has been the raw material for luxury clothing (although from a different species, *Bombyx mori*). A cocoon is made from one very long continuous strand of silk. In *Calosamia* and most other silk moths innumerable strands of silk are cemented together to form a stiff armor. It can be dented, but it is almost impossible to tear into except with scissors. One might assume that once the caterpillar is encased inside the cocoon, nothing else can enter, and that the cocoon would also trap the eventual moth inside, preventing its exit. However, without exception, in all of the many hundreds of *Calosamia* cocoons I have handled in the last seventeen years, every one had a built-in escape hatch. You can be sure that if it had been up to me to build such a hatch every time, I would have missed several. But the caterpillar does not think ahead—if it did, it too would have failures. Rather, it is behaviorally programmed, and it cannot do otherwise than leave an outward-flanging sleeve at one end of the cocoon. In contrast, the common silk moth, *Bombyx mori*, and some of the luna and polyphemus moths leave no escape hatch; instead, after molting from the pupa the adult secretes a saliva that contains a silk-digesting enzyme (cocoonase) that dissolves an exit hole out of the cocoon prison.

The *Calosamia* caterpillars have a curious behavior that is not found in any of the other local silk moth species. They spin an extension of their cocoon along the one-inch- to ten-inch-long stem (petiole) of the leaf and—more important—onto the twig beyond. This beltlike strap of silk, on reaching the twig, is wrapped in a tight ring around it. By fall the leaf wrap dries and shrivels around the cocoon, and since it is "silked" to the twig it remains on the tree even after the leaf loosens its hold on the twig. *Calosamia* cocoons may hang from trees for several years, long after their contents have emerged and long after the camouflaging leaf wrap has disintegrated. However, despite all the protection that the cocoon provides from birds, the enclosed pupa is not safe from parasitoids. They have honed their attack strategy as much as the caterpillars have honed their defense.

There are a number of different ichneumonid parasitoids (predators that kill their hosts by eating them from the inside). One of them, *Gambrus nuncius*, is tiny and prettily red-colored, with white-ringed antennae. More than forty may emerge from a single cocoon. Another parasitoid, the large yellow-orange ichneumon *Enicospilus americanus*, is relatively rare, and only one individual ever develops in any one moth pupa. Others, mainly fly parasitoids, attack the caterpillar and emerge from it before it would normally spin its cocoon.

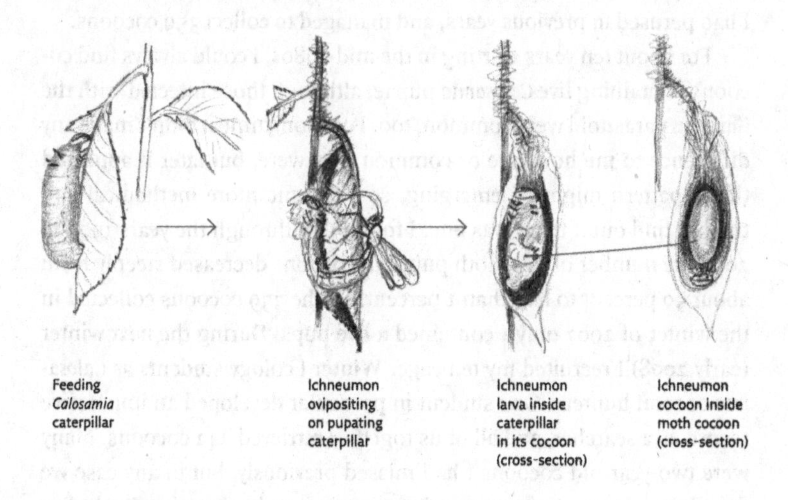


Fig. 23. ***Enicospilus americanus*** **injecting an egg into a cocoon-spinning caterpillar, and the ichneumon's development into the pupa.**

On my hill in Maine, there have usually been enough *Calosamia promethea* cocoons for me to find ten to a dozen in an hour. However, I have never seen a female moth flying free in the woods; nor have I picked up a dead one. The moth's life is restricted to about a week, and the whole population is active at the same time, in early June only. To see male moths is easier, but one needs to resort to a trick. Using a thread tied around her ample waist, I tether to a branch a female that has just emerged from her cocoon. By late afternoon males fly in, following the female's odor plumes. I collected eggs from the mated females to raise the caterpillars.

The best time to find *Calosamia*, so as to raise them from eggs in the summer, is in the winter. After the deciduous trees have shed, the few individual leaves that remain are conspicuous and may be curled around and contain a cocoon. I keep an eye out for them on every winter walk in the woods. It has not been merely an idle pastime, because I justify it by hatching females from cocoons to tether onto twigs the next summer, and by trying to find out what parasitoids the cocoon contained or still contains. I collect about 100 or 200 cocoons each winter. In the winter of 2007 I made a thorough search of the same 300-acre patch of woods that I had perused in previous years, and managed to collect 359 cocoons.

For about ten years starting in the mid-1980s, I could always find cocoons containing live *Calosamia* pupae, although those infected with the *Gambrus* parasitoid were common, too. For a long time it didn't make any difference to me how rare or common they were, but later it appeared that a pattern might be emerging, so I became more methodical and tried to find out if there was one. I found that through the years 1993 to 2006 the number of live moth pupae in cocoons decreased steeply from about 50 percent to less than 1 percent. Of the 359 cocoons collected in the winter of 2007 only 1 contained a live pupa! During the next winter (early 2008) I recruited my ten eager Winter Ecology students as *Calosamia* cocoon hunters. One student in particular developed an impressive record as a searcher, and all of us together retrieved 242 cocoons. Many were two-year-old cocoons I had missed previously, but in any case we found not a single current live *Calosamia* pupa, and only 1 contained a live parasite (*Gambrus*) pupa. In other words, the moths had now become at least locally extinct. We had collected the artifacts of a past population.

We have no idea what caused the collapse, or why other giant silk moth populations have exhibited similar spectacular recent declines.

In western Vermont near Burlington, *Calosamia* cocoons have, in contrast, been very difficult to find. All the students who hunted *Calosamia* in Maine are ardent field naturalists. One of their academic requirements—one that they find easy to fulfill—is their "Friday field walk" in the surrounding Vermont woods. For the last ten years they have been assigned to find *Calosamia* cocoons. I had found none; they found four. Three of the cocoons had successfully eclosed moths, and the fourth contained viable *Gambrus* pupae. *Calosamia* still live, but at very low densities. Perhaps the density of *Calosamia* is so low in Vermont that disease organisms and parasitoids have a hard time finding them, but the moths can still find each other to mate.

The males' female detectors are evolved to what looks like extraordinary perfection. In early June 2004, 2005, and 2006, my screen cage in Vermont containing moths from cocoons that I had collected in Maine was aflutter with freshly emerged moths. I did not release them, because I was hoping to get eggs for a crop of caterpillars. The day after they emerged, at around six PM in bright sunshine, the big magenta males "with a bolt of lightning crackling down their wings" as one moth aficionado described them, were flying all around our house. It seemed that a swarm of them had gathered. Males were again flying around the house the next day, in the late afternoon, but by the day after that I saw no more of them.

I next hoped to raise caterpillars from the eggs laid by the mated females.

Using a thread looped around their "waist" I had tied females onto both ash and chokecherry bushes, and each of these mated females deposited about 100 to 200 eggs onto the leaves and twigs there. Freshly laid eggs are sticky and become glued onto the substrate. I checked later, expecting the bushes to be crawling with caterpillars, but found not a single one. In the third summer I took the mated females into my study and let them deposit their eggs in screened cages so that I could keep track of them. All the eggs hatched, and hundreds of young caterpillars started feeding on the ash leaves I had provided for them. Then, soon after their first or second molt, they all died—every single one. Not since

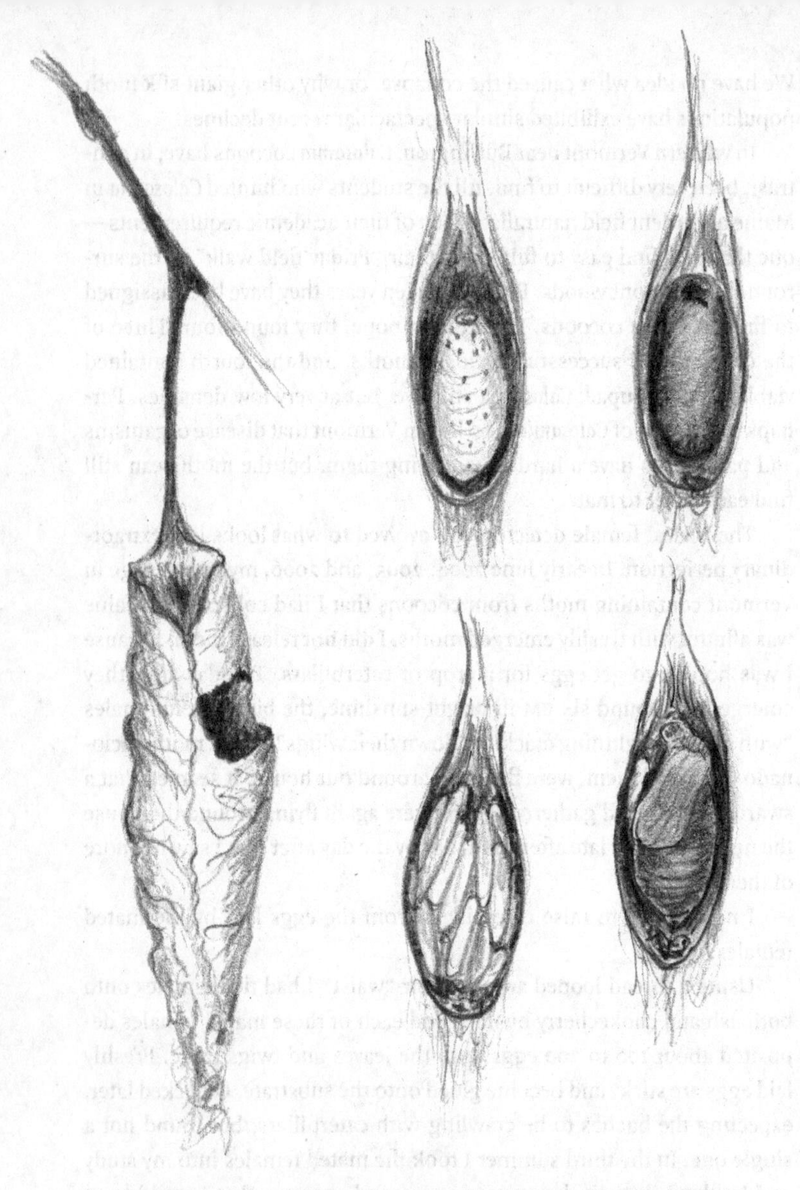

Fig. 24. **Various contents of *Calosamia* cocoons. *Left:* An empty cocoon chewed into by a rodent. *Top left:* Mummified caterpillar killed by a mold. *Top right:* Cocoon of the *Enicospilus* wasp, showing a hole at top where it emerged. *Lower left:* Cocoon packed with numerous *Gambrus* wasp cocoons. *Lower right:* Cocoon showing the pupal skin of a successfully emerged moth.**

I raised caterpillars as a small child, when a fellow student zapped my screened caterpillar cage with spray from a Flit insecticide dispenser, had I seen such caterpillar mortality. The signs this time pointed to death by a virus. I noticed also that the few gypsy moth caterpillars that I had found near the house in Vermont died as well; they turned into a liquid broth.

The gypsy moth, *Lymantria dispar*, is a forest defoliator in North America. It evolved in Europe and Asia and was introduced near Boston in 1868 by E. Leopold Trouvelot; and, as they say, "The rest is history." It became such a serious forest pest in all states east of the Mississippi that state and federal governments instigated eradication attempts, spraying millions of acres of forest with pesticides and also releasing biocontrol agents that have included viruses, bacteria, fungi, a beetle, and fly and wasp parasitoids. Some of these biocontrol agents affect the gypsy moth specifically. Unfortunately, others have a broad spectrum, attacking so-called nontarget species of native moths and butterflies.

Recent studies by George H. Boettner and Joseph S. Elkinton, entomologists at the University of Massachusetts-Amherst, have demonstrated a deadly effect of one of those gypsy moth biocontrol agents on the caterpillars of these moths. It is a parasitoid fly, *Compsilura concinnata*. This fly alone "controls" at least 200 species of moths. "It is," Boettner told me, "an evil beast that should never have been released." Are there also evil beasts among the viruses, bacteria, fungi, and introduced wasp parasitoids?

About 150 species of insects alone have been released deliberately for biological control. To test for the presence of parasitoids, Boettner regularly puts thousands of hand-reared silk moth caterpillars into the wild. Sometime later he retrieves them to study their parasite load. He has found that in some localities 100 percent of the silk moth caterpillars are infected (killed) by the one introduced fly alone, in less than one week. That's a disastrous first inning for the caterpillar in its game with just one of its many enemies, as it tries to reach adulthood. The caterpillars are normally exposed to these flies not for one week but for a month, and in order to turn into adults they have to survive not only this exposure but the pupal stage as well. Add to this mortality caused by flies, also that of caterpillar-foraging birds, and by the fungal, bacterial, and viral diseases that also affect the larvae, and it is a wonder that any ever remain to spin a cocoon.

I had, perhaps by chance, observed one very prominent moth's precipitous decline on my little plot of land on the hill in Maine. However, there are thousands of species, all invisible links in the fabric of any ecosystem. It is hard for anyone to care about any of them, because so little is known about them. And probably not much ever will be known—because there is no apparent proximal reason to know about them except curiosity for its own sake. "Every country boy" was familiar with one very conspicuous moth—the *Calosamia*—in the past, as Holland wrote in 1903. Is anyone familiar with it now? There are thousands of other more or less invisible species. Yet we don't have people able to identify even most of the big brightly colored "canaries," never mind important players like *Gambrus* wasps that are "biocontrol agents" and play a vital role. There is even debate about how many and what kinds of *Gambrus* wasps there are, and we can't afford to lose any of them.

I think the lesson of the moths is the value and power of even the rarest players in the maintenance of a natural ecosystem, and the danger that even a few members of an alien species can disrupt it. High summer temperatures allow for rapid growth, and the long growing season then permits several generations in small animals. Generation times that are speeded up to a year or less, and mortality due to predators and parasites that is then ramped up to at least 99 percent, may appear alien to us. But in such a system any small difference in an inherited trait has an extremely high chance of being "noticed" in terms of evolution. The parasitoids (and bacteria and viruses) should therefore always be "ahead" of the slow reproducers. Of course the parasite's raison d'être is not to kill. It is to live. But if its host's population is dense, then virulence is adaptive because the parasite can grow without restraint within its host, eating every bit to proximally put out the maximum number of offspring. Killing the hosts in the process does little harm to the parasite, which simply jumps onto the next host. However, once the population of hosts becomes very rare, then any parasites that kill the host quickly become extinct, and the benign parasites are selected instead. It's a case of "overexploitation of the resource." It is wildly successful—for a while. But once the resource—in this case, live hosts—become rare, then the parasite, or the bacterium, or any other infective agent will be committing suicide if it kills, because it will then die with its host. At low population the tables turn, and then only the benign parasites will live, a situation analogous to what happens when a society produces an invention that allows it to tap new resources or invade virgin territory. The exploitive strategy is favored when resources are plentiful; but when they run out or there is no place else to go, the frugal strategies then persist preferentially.

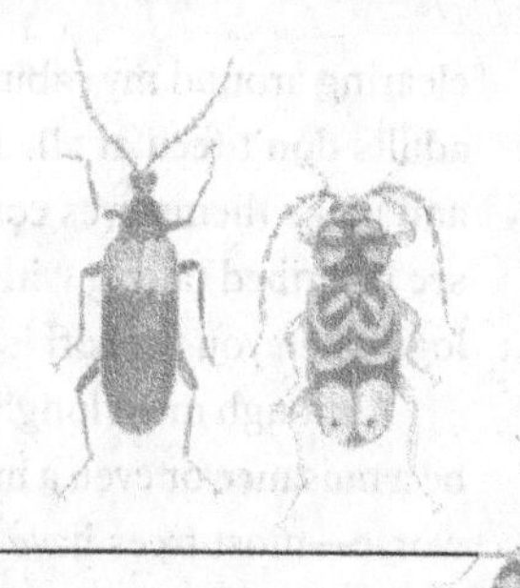

# 12

# New England Longhorns

23 July 2005. TODAY I FOUND SEVERAL FOOT-LONG LIVE oak twigs on the ground under the tree I had planted about fifteen years ago next to the cabin. How could they possibly have broken off? Looking at the breaks, I see a circular groove; they have been girdled, as though by a sharp knife. There is nobody else here, and they came off from near the top of the tree. I wonder what happened, and I sketch what I see. Is it a sign of a longhorn?

LONGHORNS ARE AN IMPRESSIVELY LARGE AND DIVERSE group, and they are not exclusive to Texas. They are certainly more interesting than those notorious longhorns that are sliced and served at restaurants. I'm talking about longhorn beetles, of course—those that go by the family name of Cerambycidae. You need a field guide to identify most of them, and my guide with color plates, by Douglas Yanega, which is restricted only to those of New England, lists 344 species. All have long curved "horns" (antennae). Body markings vary from pastel browns, grays, and black to garish yellows, blues, and orange-red. The colors are arranged in all sorts of intricate stripes and patches. I don't think I've seen more than a dozen or two dozen species, although some can't be missed. The adults in one group of species feed on flowers in the clearing around my cabin in the Maine woods. In most other species the adults don't feed at all. The larvae of many species eat bark and wood, and make themselves conspicuous by their feeding "tracks," which you see inscribed (along with those of some other insects) on the surface of logs when you peel off loose bark.

Although most longhorn beetles don't feed as adults, their larvae can be a nuisance or even a menace to trees. However, this situation could be worse—most trees have evolved defenses against longhorns, which are one of their main enemies. For example, balsam fir trees exude sticky resin when their bark is injured, and any beetle grub attempting to enter the body of the tree will be immediately challenged by a chemical counterattack of sticky, unsavory petrochemical-scented resin. In turn, the beetles have evolved a more nuanced attack. They wait until a tree is dying or very recently dead and defenseless before they lay eggs on it, and their larvae can start to eat the still moist and as yet unspoiled carcass. Indeed, longhorn beetles have an uncanny ability to detect the smell of death and injury on a tree, because invariably in the summer when I chop down a pine, fir, or spruce, one group of these beetles, the sawyers, *Monochamus*, come flying in—within minutes! Undoubtedly the beetles' chemical sensors, arrayed on their "horns" (which are a little longer than body length in females or more than twice body length in males), are attuned specifically to chemicals in pitch, and in the case of the males, presumably also to the females' scent.

The larvae that hatch from the sawyers' eggs burrow under the bark and later bore into and through the wood. Within weeks you hear their loud chomping—a common summer sound in the Maine woods, resembling that of a crosscut saw. Piles of "sawdust" (dried digested wood) accumulate under most logs. However, as far as I know, the sawyer beetle grubs are never successful in attacking a healthy tree. As a rule, longhorn beetles attack only dead or dying trees, and when they do, it's in droves. Like bark beetles attacking a fir tree, wolves attacking a moose, or male wood frogs attracting females, they succeed by cooperation even though they are proximally opportunistic competitors.

Although very few longhorns can successfully handle a whole live tree by direct attack on the trunk, some can take trees limb by limb. I was surprised, for example, when I found the one-third-inch-thick twig of a red oak tree on the ground next to my cabin. How did this get here? I wondered. It looked as if it had snapped off, but oak twigs don't snap off as neatly as the break seemed to indicate. Looking closer, I saw that the twig had been girdled. I immediately suspected a longhorn beetle larva. It seemed logical to suppose that in the summer the nutrients traveling down from the leaves through the branch were accumulating at the girdle where the larva could intercept them. But, no, I learned that girdling is the work of an adult female. In this case the female had worked long and hard for many days, chewing through solid wood, to girdle the branch. Then she had deposited an egg in the dying twig. The twig would then break off. That is, she had neutralized the tree's ability to defend its limb and provided food for her offspring. I had witnessed a week's summer work of the oak girdler, *Oncideres quercus.*

There are exceptions even to the generalization that the larvae are restricted to dead trees or parts thereof. One of them, the sugar maple borer, *Glycobius speciosus*, is a large, conspicuous beetle with bold yellow markings that mimics a yellow jacket wasp. (Not to be confused with the large black white-spotted Asian longhorn beetle, *Anoplephora glabripennis*, that is currently infecting sugar maples in New York and Chicago.) This native sugar maple borer would be hard to miss if it were to fly by or land near you, and the evidence of its presence in our New England woods is even more prominent. This beetle deposits its eggs under sugar maple bark crevices in the summer. The larva then chews into the bark and continues to chew, making a burrow in the inner bark and sapwood. Sugar maple, this longhorn's food plant, is not called rock maple for nothing, but the soft, flabby white grub manages to chew through the solid live wood with its pair of small but apparently iron jaws (mandibles), to make a chamber in the wood where it stays for the winter. It resumes feeding under the bark during the following summer—unlike most insects, for whom one summer per lifetime is enough. A baby warbler eating caterpillars can reach full size in six days, but the diet of wood eaten by the longhorn larva necessitates slow growth—it reaches full size only at the end of its second summer. The next spring, it burrows deep into the solid wood, where it excavates a cavity and leaves an exit hole for the adult to finally escape during the third summer, to complete its brief life as an adult.

Unlike the young pine sawyer grubs that feed in the inner bark of a dead pine by making random burrows, the young sugar borers often burrow horizontally in the inner bark of the upright tree. This inner bark is phloem, a live tissue that transports the photosynthetic products of the tree, principally sucrose, downward. The larva's feeding interrupts this nutrient flow and by girdling the tree, produces maximum damage. The girdling kills the wood above and below the feeding furrow of a single larva, and that furrow later leaves a huge scar on the tree, one that becomes more visible as the tree continues to grow.

Fig. 25. Feeding damage of the native sugar borer.

The destructive girdling feeding pattern could have a practical advantage for the larva. Perhaps the tree can cut off nutrient flow around a larva, but by girdling nearly all the way around the whole tree the larva is assured of a fresh food supply. Even a single larva could easily kill a tree, if it went just a little bit farther to complete a full circle all around the tree, as the oak borer always does around a twig. And now we have an irony and a puzzle: the beetles, despite their deadly power, damage but do not devastate the population of sugar maple trees.

As I have indicated, the pine sawyers converge and come a-flying, probably from miles around, to attack a single injured or dying tree, and that tree is then soon riddled with hundreds of larvae within days. In the woods of New England there are sugar maple trees of all ages, and they are one of the most dominant of forest trees. Since the sugar borers attack only healthy trees, they have a practically infinite food supply that stretches from southern Canada south to North Carolina and west to Minnesota. Any adult beetle emerging from one tree has other food trees directly adjacent to it, and it could presumably lay hundreds of eggs and summarily kill them all. Yet by far the majority of sugar maples, although this tree is the exclusive host of this beetle, are uninjured. The beetle's typical mark is found in only one of perhaps hundreds of trees. The beetle defies the predictions or extrapolations of what would almost qualify as one of the many unsavory "laws" of nature: to multiply until resources are exhausted, and then to "crash" in a massive die-off that then starts the whole process all over again.

Why does the sugar borer's population not skyrocket? Why does the borer not eat all that it can until its main resource, sugar maple trees, has been devastated? What prevents the familiar, often frightening scenario that is generally avoided only because of parasites, diseases, and predators that multiply as soon as the population increases above a critical level? Nobody knows. This is not rocket science, but it too is complicated.

The sugar borers have achieved, or are held to, something enviable. They are in a world of plenty, so none go hungry, destroy their habitat, or jostle and interfere with each other. Somewhere there is a check on their natural rate of increase, and you can be sure of one thing—that if they could tell us what they wanted at any one time, they would vote to obliterate the forces that hold them in check, the forces that ensure their long-term benefits. And so, probably, would we, if we voted merely on the basis of our individual interests.

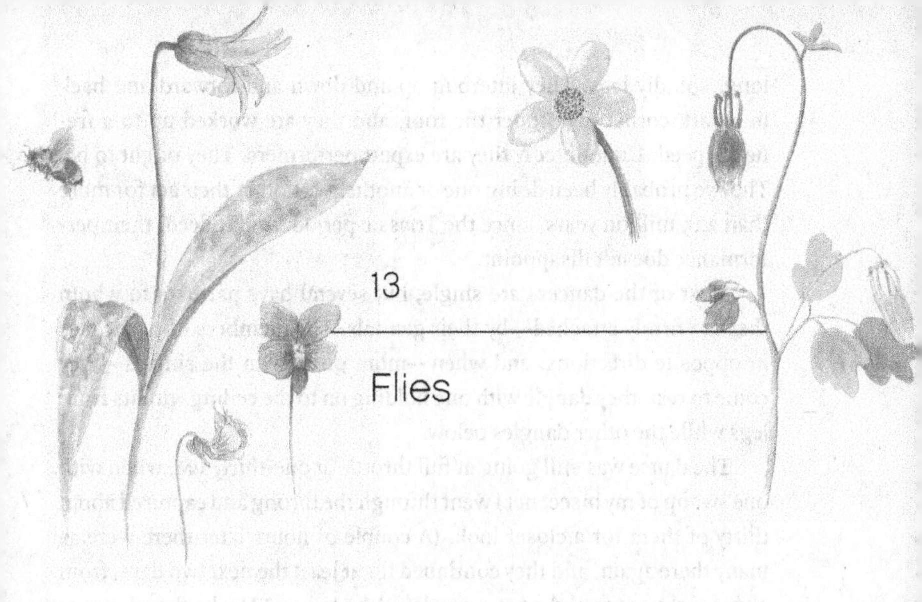

# 13

# Flies

We breed 'em, you feed 'em.

—BUMPER STICKER OF
THE MAINE BLACKFLY BREEDERS ASSOCIATION

21 June 2007. IT'S THE SUMMER SOLSTICE (IN THE NORTHern hemisphere), and according to my calendar, which only one species goes by, it's the "first day of summer." But for many species summer has actually been in progress for months, and it's now arguably the middle of summer—when the days are longest as the Earth's axis tilts the most toward the sun. The hottest days, though, are still to come. In any case, it's sufficient reason to celebrate, and what better way than to enjoy a dance performance?

As chance would have it, I find one. It's right here at my camp in the Maine woods. The dance is in the outhouse, presented by a special troupe of untiring performers. I'm just a spectator today, and viewing conditions are perfect. It's a pleasant 70°F—too cool for horse and deer flies and too dry for blackflies and the god-awful midges, the scourge from hell.

Our outhouse is open at the front, and it faces deeply shaded sugar maple woods. The dancers—two or three dozen of them—each have six long, spindly legs. They jitterbug up and down and forward and back in a dark corner just under the roof, and they are worked up to a frenetic speed. Undoubtedly they are expert performers. They ought to be. They've probably been doing one or another version of their act for more than 225 million years, since the Triassic period. And indeed, their performance doesn't disappoint.

Most of the dancers are single, but several have partners to whom they are firmly attached—by their genitals. The members of a pair face in opposite directions, and when—more often than the singles—they come to rest, they dangle with one holding on to the ceiling with its front legs while the other dangles below.

The dance was still going at full throttle at one-thirty PM, when with one swoop of my insect net I went through the throng and captured about thirty of them for a closer look. (A couple of hours later there were as many there again, and they continued for at least the next two days, from about eight AM to eight PM every day. Who knows? Maybe they dance at night, too.)

Superficially they resembled huge mosquitoes. They are relatives of this group of insects, commonly known as crane flies because of their very long legs. Their bodies were about a third of an inch, while their legs were three times longer. Their legs drop off at the merest touch, an adaptation for making a quick getaway from a predator. But these didn't get away, even as my sweep of the net left the bottom of it littered with a small pile of loose legs. These flies all looked similar, except for their genitalia. Of a pair I examined, one had a thicker but pointed abdomen, and the other a thinner abdomen with a blunt end. I presume the more ample individual was the female. Further examination under a magnifying glass revealed a tonglike clasper at the business end of the male.

By far the majority of performers at this mating dance were males. In my sample of thirty, males outnumbered females twenty-eight to two. The dance is done primarily by single males, and the females are, as among wood frogs, presumably attracted by the communal male display. A male finds a female (or vice versa), and then they mate and leave the crowd. Once a male snags a female he tends to go steady with her, for a while. To test their fidelity to each other I put three couples that I had captured in a jar. One of these pairs, which had been on the ceiling in the early morning (presumably since the night before), separated instantly.

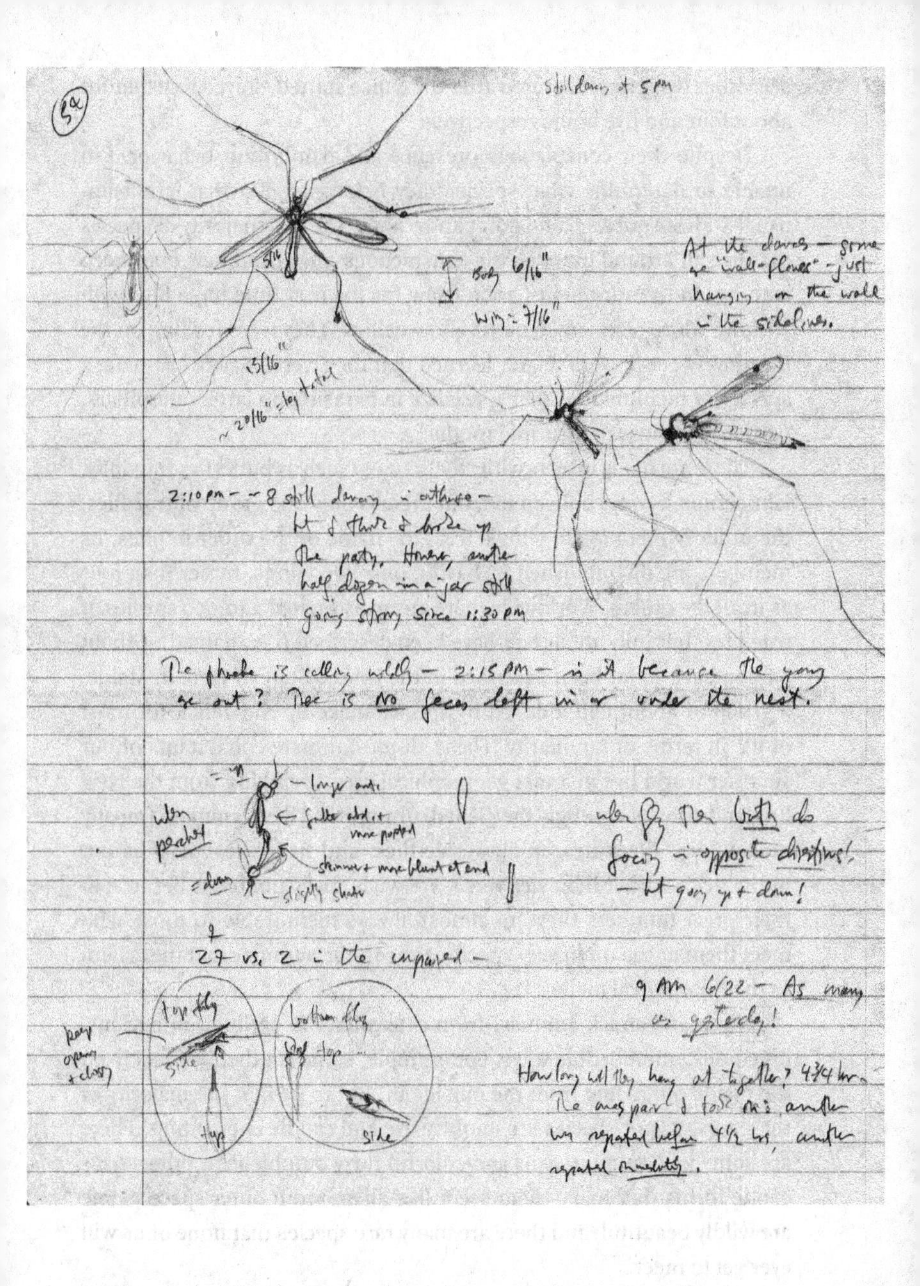


Fig. 26. **Preliminary notes on and sketches of crane flies.**

The other two pairs, captured after the dance started, stayed together for about four and five hours respectively.

Despite their conspicuous presence and flamboyant behavior, I'm unable to determine what species they belong to. But that is not unusual—I have not seen and don't know about the vast majority of species that live all around us, even the conspicuous ones. And new ones keep coming. In the summer of 2006 I saw for the first time huge flies with striking white faces and lemon yellow bellies. They were feeding on the meadowsweet flowers. I later learned that they were *Belvosia bifasciata*, a species of tachinid flies that specialize in parasitizing large caterpillars, particularly those of saturniid moths.

There are many insects with "fly" names (such as butterfly, dragonfly, ichneumon fly, and dobson fly), but there is only one group of true flies, the order Diptera ("two wings"). As the name of the order implies, its members are distinguished by having only two wings, rather than four as in all the others. Worldwide there are an estimated 240,000 species of true flies, but only about half have been described (i.e., named). (About 25,000 have so far been described in the United States.) Here in Maine, a group of about half a dozen fly species make up the deficit for most of us, in terms of familiarity. These all-too-intimate cohabitants of our summer world live in a vast geographical area stretching from the New England forests through the Canadian tundra. These animals (mostly mosquitoes, blackflies, midges, deerflies, and horseflies) seek us out in the flesh, rather than vice versa. We wish they would not. Because of their great numbers they are almost always memorable to those who meet them at the often very specific time in the summer that they claim as their ecological niche.

Dipterans attack animals, from caterpillars to caribou, in devious, ingenious, and horrible ways. For example, some eat their victims from the inside out, some from the outside in. But to be fair, the majority of the thousands of species are unobtrusive and can be enchanting. There are some who mimic wasps and colorful furry bumblebees; others have exotic forms that make them seem like aliens from outer space. Some are wildly beautiful, and there are many rare species that none of us will ever get to meet.

Every fly has its place and its season, and many flies have a specific time of day (or night) when they are active. Those that I know most intimately are not as entertaining as the dancing crane flies (there are also hundreds of species of these), who must remain anonymous for now, as I do not know their names. With regard to the next lot, who are familiar in summer, I shall be scarcely more specific and give only generic names.

Mosquitoes, the first on the list of familiar local dipterans, are the least objectionable, because, unlike species in the tropics, the northern species are not carriers of malaria, dengue fever, yellow fever, or other diseases, as far as we know. Female mosquitoes suck blood in order to get enough protein to make a few hundred eggs, which they deposit in water. The aquatic larvae filter-feed on microscopic particles. They breathe air and, swimming by a wiggling motion, they periodically come up for air, which they suck in through a short tube at their hind end. Almost any pool of standing water where there are no fish is alive with these "wigglers" in the summer. Of course at this stage they are no bother to anyone. It is after they eclose ("hatch") from their floating pupae that they become bloodthirsty fiends. Those we meet tend to be mostly the females; the males fly around in search of flowers for nectar, females for mating, or both.

Because I'm a mostly diurnal animal and like sunny, open vistas, I have never been much bothered by our mosquitoes. Arctic ones are a different matter altogether, as for some reason mosquitoes get fiercer and more numerous the farther north you go. Caribou may become so depleted of blood by millions of teeming mosquitoes that they lose weight even while grazing full-time.

The only animals that consistently prey on mosquitoes are dragonflies, and they have probably been doing so for at least 100 million years. Mosquitoes seem to have habits that are designed to avoid overlapping with these predators. They avoid the sunshine, where dragonflies are most active, but hordes of mosquitoes appeared as soon as I stepped into dense shady woods where there were no dragonflies. That is, wherever and whenever dragonflies were scarce, mosquitoes were abundant.

Dragonflies that fly at dusk can cash in on mosquitoes. I suspect that the dragonflies' extraordinary eyes developed to keep up with prey trying to escape into the dark. Behavioral adaptations have the same effect. While walking in grass in Botswana during the heat of the day, I saw mosquitoes spring up, and I was then followed by several dragonflies that were hawking them. The dragonflies seemed to be following me directly, because when I shifted to a slow jog they continued to follow me. They were acting like some species of birds—cowbirds in North America and cattle egrets in Africa—which also follow large animals because of the prey these animals flush.

Dragonflies are opportunistic. On the evening of 23 July 2005 at about eight o'clock, the air all the way down from our lawn to the beaver bog was full of huge dragonflies. Hundreds were visible, zigzagging back and forth—fairly low, about ten to fifteen feet above the ground. I had never before seen so many at once. Ten minutes later—the sun was still five degrees above the horizon and the honeybees were still working on the flowers—the dragonflies were suddenly done flying.

Mosquitoes are at times an irritation almost anywhere, to be sure, but by knowing their schedules one can avoid many of them. Mostly, this means staying indoors at night. The mosquitoes who get me while I sleep leave a tiny welt that disappears in a few minutes. Newcomers to the woods, who have not yet paid their "entry fee" to nature, don't always get off quite so easy. It is not a good idea for bare buttocks to be exposed after dark, especially if one's immune system has not yet been fine-tuned to receive their attention. Big red welts that itch to distraction are a consequence.

Blackflies fill in, and then some, where mosquitoes leave off. They are any of a number of species of small (about 0.08-inch) hunchback flies with thick stubby legs (all the better for crawling into your hair, and through creases and holes under your clothes), of the genus *Simulium*. Their larvae are filter-feeders that attach themselves to rocks at the bottom of swiftly flowing streams, often in such numbers that the rocks look coated with black mats of moss—but each "moss" frond is a larva. These larval mats may extend endlessly in a stream, and when the adults emerge half of them are hungry for blood. (The other half will be males who don't need the protein meal.) The flies are silent—unlike mosquitoes, they emit no hum on their approach—and as soon as one surreptitiously lands, it starts sawing into the flesh. Blackflies drive moose to distraction, and affect some people even more severely. People who live near a wilderness or venture into the woods in "blackfly season" (i.e., summer), even those who have over the years developed an immune response, consider these flies quite a nuisance.

My first memories of Maine blackflies are associated with trout streams at a time when my backwoods mentor, Phil Potter, tried to make a man out of me—and out of his young nephew, Bertie. I don't recall how well he succeeded with either of us, but I won't forget our simultaneous entanglements with trout fishing lines and blackflies in the alder bushes where we were wading in cold water, while the part of us above water level was all lathered up with DEET. No matter what, the blackflies always managed to find entry points along the sleeves, collar, hair, fly, nose, mouth, and ears. Especially the ears.

My most memorable incident with blackflies occurred in Ontario. My wife and I, our young daughter, and our dog Foonman were making our annual trip to Maine from California. It was a warm, humid summer day. We stopped off in the woods to let the dog out for a brief romp. He jumped out of the car and headed for the nearest tree to lift his leg, but his pit stop was uncharacteristically brief. He raced back to the car even more eagerly than he had left it, chased by a diffuse black cloud.

The problem for us is that blackflies are diurnal. They are active at about the same time as we are and in the same places where we like to enjoy the summer world—out in the woods, the garden, or the trout stream. Few people who have not experienced blackflies have any idea what they can be like. At least I assume so, because I often hear someone complain bitterly about blackflies where they are practically nil. I think, "What blackflies?" To see *real* blackflies you need to check out the northern woods on any warm day between 1 May and the second week of July. The rest of the summer is usually relatively free of them.

It is possible to somewhat reduce the flies' depredation, by knowing their habits. First, unlike mosquitoes, blackflies avoid dark, enclosed spaces. Therefore, they don't molest you in the house, even if the door is kept open. And timing is crucial; they can be almost totally absent on a clear morning, if it is cool. They are equally reluctant to fly at high temperatures: those above 85°F are the niche claimed by another group of flies, the tabanids.

The tabanids, high-temperature specialists, are large, compact animals with short legs and a fast, quiet flight. They have huge, commonly iridescent green eyes, and some of these are colloquially known as "copperheads." The group includes deerflies, moose flies, and horseflies. Like blackflies, these do not just pierce your skin; nor do they "bite."

Instead, their mouthparts have a bloodletting tool consisting of two side-by-side scalpels that, working like alternately moving blades of a pair of scissors, cut into the skin. While they are at it, like other bloodsuckers and blood lappers the world over, they also rub it in by simultaneously applying anticoagulant from their spittle. They make the blood flow and keep it flowing so as to lap it up all the more easily. Unlike the blackflies, these large flies—in the case of some species, huge flies—are often conspicuous as they zoom around you, yet they are often very cautious in landing. You anticipate a wallop of a "bite" (an incision) at any second, but the critical and continually anticipated event may be delayed for minutes as they keep circling around your head, waiting for an opening. While I'm running, they fly circles around me, searching for an opening, which is usually under the damp hair on the back of my head.

Of the many biting flies, for me the most objectionable are the smallest: the midges, also called no-see-ums. They are especially active on warm, balmy nights. Not only don't you see them; you don't hear or smell them either. But when they arrive, you know it. You feel crawling, burning sensations on all exposed areas of the body, and then also on the unexposed areas. These insects come at you even inside your dwelling, and most window screens are no obstacle to them.

A friend, a guide in Maine, once entertained a couple of summer "sportsmen" from New Jersey who had never encountered midges before. He told me of one memorable incident with them. The party of three arrived at dusk at their bucolic, Edenic campsite deep in the Maine woods. As they were unpacking and getting set up, the sportsmen began to scratch. Then, suddenly realizing that they were under attack, they jumped up and went "literally crazy," as my friend put it. They eventually ran off into the woods, trying to escape their tormentors. Unlike the superfast tabanids, which not even a deer can outrun, no-see-ums can be outrun by a reasonably fit man. But the problem is that there is no place to run to, because they are everywhere.

Smoke. DEET. Running. Nothing worked to diminish the sportsmen's pain. Even sooner than anticipated, they resorted to trying to distract themselves with their precious stock of beer. However, the night was hot and muggy and the midge onslaught was long; their beer supply ran short. After that my friend had to spend the night with his guests in their truck, driving back and forth on a bumpy road through the forest to create a cooling breeze that would blow the midges off. It worked, until their gasoline ran low. Drained in more ways than one, the summer vacationers sped back to New Jersey in the morning.

Well, at least in New England nobody has to pay homage to botflies. These are nonbiting flies, but they can be even more bothersome than the blood specialists—especially for caribou in the arctic. The large botflies bother caribou to distraction by flying up their nostrils to deposit, not eggs, but live maggots that will burrow in and wander around in the body before lodging under the skin to grow there to adulthood. When the maggots are well fed and fully grown they pop out of the caribou's skin to pupate on the ground. In winter, on a freshly skinned caribou hide, I have seen dozens of large white welts, each containing a big botfly maggot. I have also seen botflies on skinned mice and chipmunks in Maine; relative to the size of their hosts, one of these maggots would be as big as a woodchuck to us.

I've learned a few things from flies. I've learned that it's unproductive to swat gnats. I've learned that it's a good idea to look at flies carefully, to distinguish the bothersome from the benign. The good ones, for me, are those who dance for their own pleasure. I do not disdain those who suck my blood so that that they can lay their own precious eggs; they are just programmed that way. It's pointless to try to reason them out of it. It's better to take hits without flinching, and to develop an immunity to the toxins.

Aside from that, flies give me hope. They also inspire others, as I learned from a bumper sticker that made my day a little over a year ago. It said in bold black letters: "Save the Blackfly." I trailed behind the car with the sticker for about twenty miles before it finally pulled over in Plainsfield, Vermont. I pulled in right behind. I could now read the fine print at the bottom of the sticker. It said: "Maine Blackfly Breeders Association." I wanted to belong, knowing that blackflies effectively do more to fulfill the promise of the well-known state slogan to "Keep Maine green" than anything government ever would or could do to keep "development" at bay. But as I walked over to introduce myself to the gentleman in the car, he stepped on the gas pedal and quickly drove off.

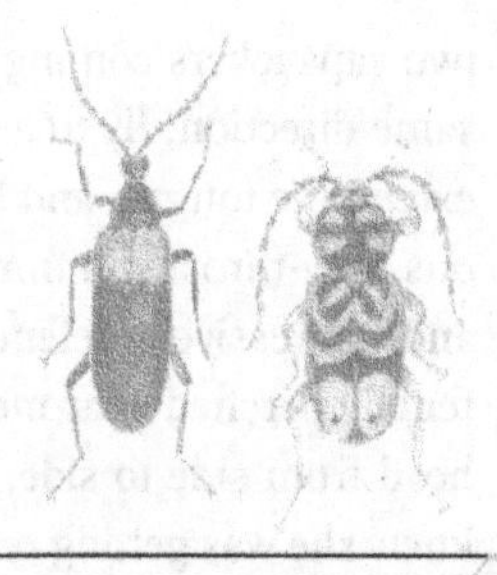

# 14

# The Hummingbird and the Woodpecker

14 April 2006. I AM UP AT SIX AM, RISING WITH THE SUN and beating it only slightly. Perfect timing, because now is when the action is. The sapsuckers have just returned, and one has found the aluminum ladder I had put up only yesterday afternoon by the shed at our house in Vermont. He is drumming on it—"rat-tatatattat-tat" over and over again. Three more male sapsuckers came to the ladder as well and I wondered if they would join in, but they chased each other instead and then all three left. One came back to the ladder several times later to drum some more. A unique bird, this species. It's the only one of the six local woodpeckers that has a non-uniform drumming rhythm, and that taps trees for sap.

17 June 2005. I'm on a platform of boards I have built about twenty feet up, between some close-spaced young red maple trees near my camp in Maine. I'm facing a single big birch tree that has four active sapsucker licks. The tree has just leafed out, although many of the other forest trees are still bare. There is a light rain, and temperatures are in the mid-fifties Fahrenheit.

Sapsucker licks are hubs of life in these woods at this time, and I want to take the pulse of this life. I hope to see surprises, even as I want to learn the routine. Today I've already established that there are at least two sapsuckers coming to this one tree. They regularly come from the same direction, fly to a tap, touch the sap hole with the tip of the bill, extend the tongue, and lick sap as they vibrate the head rapidly. Numerous ruby-throated hummingbirds come and go, and some of them stay and aggressively defend the site against others that come near. One female perched near me and about ten feet from a lick. She jerked her head from side to side, steadily, like a metronome, once per second. I knew she was getting ready for another feed at a lick when she turned in that direction, stretched, and then zipped over to it and hovered to feed for about thirty seconds before zooming off into the woods. Another then took her place. In the afternoon, a sharp-shinned hawk shot past me like an arrow through still leafless red maple branches. It landed on a limb next to a lick; nervously turned its head here and there, probably looking for prey; and then zoomed on as quickly as it had come.

22 August 2005. I could not have chosen a better day to be in the woods, up on the same sapsucker lick viewing platform. It's seven-thirty AM and the sun is up. I'm twenty feet up, under the leafy canopy of young red maples. I have brought along a hot cup of coffee from the cabin, and sipping leisurely, face the thick birch tree with the woodpeckers' licks. A few red maple leaves around me show the first twinges of red, and the field next to me is ablaze with goldenrod bloom. A flock of about a dozen chickadees come sauntering by. After they pass I hear warblers cheeping in the distance. Warblers sang lustily months earlier when they were separated into their respective territories, each in its own specific habitat. Now, all is different. A loose flock of them soon drifts in, and it includes at least eight species: one chestnut-sided warbler, several black-and-white warblers, several blackburnian warblers, several ovenbirds, many Canada warblers, many black-throated blue warblers, a couple of black-throated green warblers, and several redstarts.

Along with the warblers are at least one wood thrush, one least flycatcher, and a phoebe. They stayed busily foraging all around me, for about twenty minutes. Many of them came within several feet of my face.

Six licks (each with dozens of sapsucker holes) had been heavily used for almost three months. But in one hour of watching I saw only one hummingbird come for a sip. This female (or juvenile) flew to all of the six taps but stopped only briefly, as though inspecting them and finding them dry. Except for one or two black-throated blue warblers, which hovered twice in front of a sapsucker lick but did not contact it, none of the warblers paid any attention to them. Only one sapsucker came, a juvenile from this year's hatch. It perched about six feet from a lick for a full twenty minutes; moved its head occasionally to scan around; then stretched, preened for another ten minutes, and finally briefly inspected two licks before departing.

Twice an admiral butterfly flew by the licks without landing, and the juvenile sapsucker made a pass at it but missed. A knot of a dozen bald-faced hornets aggregated at one spot at one lick only. Perhaps they had located the one last bit of sap flow. A downy woodpecker, which had previously sometimes used the same lick, came within a foot of it, looked at it, and then flew back into the forest of mostly maples, pines, and firs. It seemed to me as though the steady customers at this lick site were coming to check on it, but not finding much of anything.

I come down from my perch and then climb up the birch to inspect the licks. As expected, they are indeed dry now although the woods are still lush and green. But what a difference from what they were like a few weeks earlier! There is no more birdsong. Even the sapsuckers, who were noisily drumming all around this lick when they first came in mid-April, and who continued vigorously into late June and July when their young fledged, have now ceased. The summer is ending and life is clearly on a much more leisurely schedule. The birds may already be fattening, preparing to leave the summer world behind them.

LIFE FEEDS ON OTHER LIFE, AND IT'S NOT ALL ABOUT FLIES sucking blood. Indeed, most of the interesting and—to us—uplifting nature stories emerge from other feeding relationships. Even as I write on 2 July 2007, the *Amelanchier* (Juneberry) tree in front of my office is being visited by robins, cedar waxwings, purple finches, rose-breasted grosbeaks, a catbird, and a veery. They are feeding on the now ripening (but not yet ripe) berries, and will be spreading seeds in all directions to potentially "plant" more trees. Whether or not we approve of any feeding relationship generally depends on whether we are on the giving or the receiving end. The ruby-throated hummingbird that brightens our summer days is at both the giving and the receiving end of feeding relationships: the receiving end is from us and sapsuckers, and the giving end is the pollination of plants in Central America.

There are at least eighty-six species of hummingbirds in Brazil alone, of 343 species documented so far. America is the hummingbirds' home, and the South American tropics are the heartland of these jewels of the bird world. Here, in New England, we are graced by only one species, the ruby-throated hummingbird, *Archilochus colubris*. It is a suburban favorite that is easily attracted to a bottle of plain sugar water marked with a red artificial flower.

Like all hummingbirds, the ruby-throat belongs to a family exquisitely adapted to feed on the nectar of flowers, and many flowers have coadapted to be pollinated by them; one depends on the other. "Hummingbird flowers" have a long tubular neck (corolla) that excludes "nectar thieves"—those animals that feed from these flowers but do not pollinate them. Hummingbird flowers are commonly red. In contrast, the flowers pollinated by sphinx moths—the nocturnal analogues of hummingbirds—are white and strongly scented, to be more easily located by prospective pollinators.

Many hummingbird species are, in turn, adapted to specific kinds of flowers. Even the length and curvature of their bills fit specific flowers (and exclude others). But the ruby-throated hummingbird is, in many ways, one of a kind. It is the only hummingbird that has extended its range into the coniferous forests of northern Maine and Canada, in areas where all summer long there are only small white nectarless flowers on the mossy ground, and no red flowers at all. Farther south I had been astounded to see ruby-throats when there was still snow on the ground, long before any flowers opened, and before there were any leaves on the trees. Once in April when there was still snow on the ground, a hummingbird appeared as if out of nowhere and hovered briefly in front of my face. A friend told me of seeing another one hovering around the head of a pileated woodpecker. I had been wearing a red hat. The woodpecker has bright red feathers on its head. I suspect that these hummingbirds had recently come back from their tropical wintering grounds, where they had been feeding at red flowers. On coming north they were still responding to the same signal that had meant food. But why and how did and could they have risked leaving their tropical paradise with red flowers, to come so far north at a time when there is no red and there are no flowers with nectar?

A male ruby-throat weighs about one-tenth of an ounce, only slightly more than a penny (about 0.08 ounce) and several times lighter than many large moth caterpillars. Its heart and wings beat at twenty-one and sixty times per second, respectively, while it flies north on a journey of about 2,000 miles and then flies the same journey south in the fall. At any given moment it is within hours of starvation, as it needs to consume about twice its body weight of food per day. No other hummingbird attempts what would appear to be a very risky journey, to a destination often devoid of nectar-bearing flowers. When hummingbirds are not traveling, their body temperature often drops until they are torpid, to conserve energy when they can't forage, as at night. They prepare for their long nocturnal fast by fattening up during the day. During migration, they replenish in the morning, leaving the middle of the day for traveling. Indeed, at Hawk Mountain in Pennsylvania in the fall, most of the migrant hummingbirds come through at midday (Willimont et al. 1988). But apparently they change their strategy on another leg of their migratory route, when they cross the Gulf of Mexico.

The Gulf of Mexico presents 520 miles of open water that must be crossed at one go and without refueling. At their top sustained flight speed of about thirty miles per hour, they face a nonstop flight of about seventeen hours. And here, before tackling that distance, the hummers depart from Alabama in late afternoon for a night flight directly across the gulf (Robinson et al. 1996). In the spring the birds returning to Alabama (at the banding station at Fort Morgan) arrive in the dark of night (Sargent 1999). Apparently some of them also take the longer route along the coast of Texas, where they could migrate by short hops and presumably refuel. Do they decide on one or the other option on the basis of their fat reserves? Do they know what they are up against before they take off over the open water of the gulf?

Although migration is hazardous, it can't be excessively so for the hummingbirds, since they have one of the lowest reproductive rates of any northern bird migrants. They raise only one clutch of two young per year (perhaps because the female alone does all the work). By contrast, a pair of northern warblers will raise four to five young in a clutch, and a pair of golden-crowned kinglets will raise eight to twelve chicks per brood and nest twice per summer. Since on average these bird populations are stable over time, the number of offspring they raise provides a measure of their mortality rates; the hummingbirds must therefore have a relatively low death rate. We know they come north to build their tiny nest cups of lichens held together by spiderwebs, where the female rears her clutch of two young. But why not do it in the south? Why not stay in their ancestral home along with most others of their group? There are many theories, but no answers. There are, however, some answers regarding how they get by once they reach the north.

The male hummingbirds are the vanguard in the northward migration, as is true of most other bird migrants (Stichter 2004). The standard explanation for this phenomenon is that the males compete to establish territories in order to attract better females. It sounds reasonable. But then in the migration south at the end of the breeding season, the males *again* precede the females and the young of the current year.

For many years I thought that the hummingbirds I had seen in the northern Maine spruce fir woods in May must have returned much too early by mistake. Later I learned they were right on schedule. Their timing is synchronized with the return of a woodpecker, the yellow-bellied sapsucker, *Sphyrapicus varius*, which is their main food provider then, and all summer long.

The sapsucker is one of the most visually striking of birds. Its bold black and white markings contrast with a scarlet crown, a scarlet throat in males, and a soft lemon yellow–tinged belly. Sapsuckers differ from other woodpeckers in not having a long pointed tongue with barbs at the end. Instead, their tongue is much shorter, and the end is brushlike—an adaptation like the hummingbirds' tongue, which is also for sweeping up liquid as with a wick.

Like hummers, the sapsucker males precede the females north, and as soon as the males arrive the woods around our house resound with their calls and their drumming. Females come a few days later, and in two weeks nest holes are drilled and egg laying begins. One of these early-arriving females hit our window and was killed, so I examined and sketched her.

These woodpeckers do not excavate insect larvae from wood as other woodpeckers do, and they don't need the long tongues of other wood-

Fig. 27. A sapsucker killed by flying into a window in April 2006. It was an immature female, as shown by the ovary with only undeveloped eggs. For woodpeckers, this species has an unusually short tongue, and relative to resident birds it has very long wings (like most migrants).

peckers to explore tunnels made by longhorned beetle larvae. Nor do they excavate any hardwood, except their nest holes. They usually choose poplars that are softened by fungus (*Fomes igniarius var. populinus*) rotting them on the inside.

Adult sapsuckers eat sugar at their licks, as well as ants and other insects that also come for the sugar. They make holes through the bark, and then use their brushy tongue to lap up the sap. The most conspicuous sap licks are those on birch trees. The whole tree is commonly ringed by tiers of holes, which are visible from afar. Every year, for several years, the birds make new taps directly above and to the side of the old ones. Then the tree dies, and the birds attack another tree. On my hill in Maine I found half a dozen sap stations on birches where sapsuckers were active from late May through most of the summer. Although there were hundreds of holes in the bark at each of six lick sites on one large white birch tree that I examined, only a the topmost holes at any one lick site yielded sap—all the lower ones were dry. I lapped the sap and it tasted sweet. The sugar concentration (measured with a brewer's refractometer) was 17 to 18 percent—similar to that of concentrated nectar.

Sapsuckers come back as early as the first week of April, and it seemed a mystery to me why birch, which produces sweet sap and seemed to be so much preferred in the summer, was not visited in the spring. Nor were the birds at any other sap licks that I could see, until at least a month later. What did they feed on when they came back before summer got started?

It was not until the spring of 2006 that I figured out the woodpeckers' solution for obtaining food on their early return from migration. I had underestimated the birds' sophistication. That spring I deliberately followed sapsuckers to see what they do. To my great surprise they were tapping sugar maple trees! It had always seemed to me that sugar maples should be an obvious choice for them. But I had not seen on these trees the patchwork of holes that is typical of their work on birches, which is what I was looking for.

It turned out that on sugar maples they punched only tiny holes, here and there. These holes were almost invisible, except on close inspection, and they quickly healed. However, in early spring, and only then, any tiny hole in a sugar maple "bled" sap profusely. Any maple tree I chose to puncture with the tip of my knife yielded the same result—huge droplets of sap welling up within seconds. But damage to the bark alone had no effect; it was necessary to puncture through it, if only slightly into the wood (i.e., xylem). Here was the answer: the woodpeckers were indeed tapping maples (and other trees), but the effect on the trees was so slight then that it left almost no visible marks. Then, in summer, when the woodpeckers switched over to birches, they made huge patches of holes that eventually killed the trees. (I later found that they only *preferred* to tap the phloem sap from birches, rather than maples; I eventually found a phloem tap each on a sugar maple and a red maple tree that were used in the summer. Both looked identical to the phloem taps that are very common on birches.)

The reason for the bird's change of behavior from spring to summer depended on tree physiology and, unlike the sapsuckers, I had not considered the fluid mechanics of the trees' "plumbing." The early-arriving sapsuckers were accessing the sugar maple sap that was going *up* in the wood (xylem) of the tree, when it is apparently under pressure and comes out through any tiny puncture. Later in the season—and only then—when the xylem flow is low, and when our own traditional maple sugaring stops, they start taking sap from phloem in birches. The phloem, a living layer of the inner bark, transports sucrose and other products of the leaves *down* to the trunk and the roots. So the birches did not yield the rich phloem sap until they had put on their leaves, when they were transporting nutrients down.

On a birch, the sapsuckers make a large aggregation of wounds in a very distinctive pattern. Each individual wound is a somewhat square quarter-inch to half-inch bare patch where the bark is removed down to but never into the wood. Series of these bare patches are placed neatly one above the other in vertical rows, and many such vertical rows are created simultaneously side by side, all around the trunk of the tree, girdling it. Any one such lick site lasts for only three or four years before all of its phloem flow has been interrupted and the tree dies.

Fluid mechanics also explains why the sapsuckers end up making huge destructive lick sites on birches. When a sapsucker takes out a patch of bark on a birch tree in the summer after the leaves are on, the sap oozes down from above and eventually gets routed around the wound after the tree responds by plugging the open phloem channels. There is then still just as much phloem sap that descends down the trunk of the tree, and the bird can tap it as before by simply opening another patch of bark *above* where the tree had attempted to stanch the hemorrhage. Alternatively, more fluid is rerouted laterally, immediately adjacent to the original hole. The woodpecker may also take the next patch of bark out there, and so the area of little bare patches expands both vertically and laterally, gradually encircling the whole tree trunk as the woodpecker keeps a step ahead of the tree's defense. This results in a band of dead tissue on the tree—a situation analogous to putting more and more stones into the middle of a stream to try to stop the flow, which isn't stopped but instead, for a while, actually increases laterally. It pays the woodpecker to keep working on the same tree as the phloem flow becomes more narrowly channeled. Eventually, however, the next few taps will cut the flow off entirely, and then the tree dies. Luckily for the woodpeckers, they can simply drill into the next tree.

The isolated sap licks on birch trees are magnets for life during the summer. In 2004, when the woodpeckers established a lick station in a large birch near my cabin, I built a platform of boards in a maple tree adjacent to it, at about the same level twenty feet up. I sat there often to

Fig. 28a. **Juvenile sapsucker at a typical late-summer phloem sap lick, on birch.**

Fig. 28b. **Typical sapsucker xylem lick in spring, on a sugar maple. Note the tap hole at top center (current year) and the two holes at lower right (previous year).**

watch. In one perhaps typical watch (six to seven AM on 7 July 2005), when I again visited my sapsucker lick to "take its pulse" I saw five bald-faced hornets at one time. All these wasps were usually grouped at the same little square sap hole lick on one branch. Four red squirrels came for their sweets. I saw nineteen hummingbird visits (apparently, all by females or fledged young). There were eleven sapsucker visits, with up to four adults present, once, at the same time. I heard the contrasts of the resonant drone of the hornet queens versus the high whine of their smaller workers, as well as the deep hum of the ruby-throats; and also the sapsucker's drumming, which seemed incongruous in the summer.

Most of this drumming—often ten to fifteen minutes at a stretch—came from the same areas and occurred shortly after light at dawn and in the half hour before dark. No other woodpeckers were drumming then. But here they drummed—one to the west of me, one to the north, with a third joining in from the east. The sapsuckers' drumming at this time, after the young had fledged, was curious. These birds had long since staked out territories and had already paired up in May. Were they making territorial claims on the sap station?

I am glad I got to know the sapsuckers a little better than I had known them before. Now they had become unique. It seemed that I had discovered new neighbors. And through them I had met a diversity of life at our common summer "watering place." My summer woods are richer now than they were before.

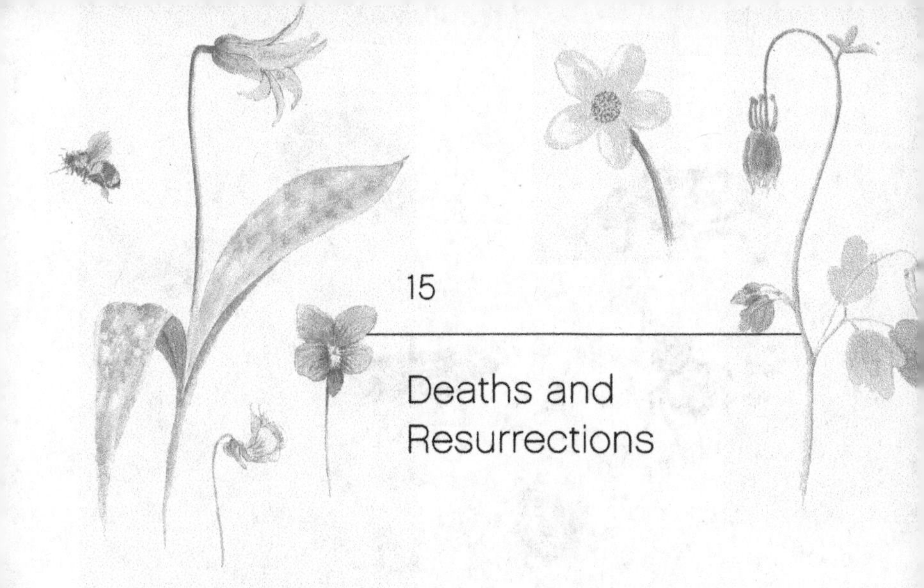

# 15

# Deaths and Resurrections

5 August 2006. I'M MET BY THE SMELL OF ROTTING flesh, and it's not hard to find the source—the remains of a half-grown wild turkey that had been killed and partially eaten by a coyote or a hawk. The coyotes here in these Maine woods are nocturnal, and the turkey had been killed next to where it was taking a dust bath on an anthill by my maple grove, so it was killed in the daytime. A coyote would have dragged it off; maybe it was killed by one of the pair of red-tailed hawks living in the area. I lift the carcass and find meat left on it. To my surprise I also find a horde of hundreds of shiny black beetles, which scuttle off and burrow into the duff of dead grass and decaying leaves. They are shiny, streamlined, and fast. I dig after them with a stick, and discover also two species of boldly marked orange-and-black sexton (or burying) beetles. They tuck in their legs and play possum as soon as I expose them. These beetles are monogamous and care for their young, which they rear in a small nest. The parents gather meat and, in response to their grubs' begging, regurgitate the half-digested food to them. The father repels intruders, mainly other male sexton beetles that try to kill the babies and try to mate with the female to get her to produce a second clutch, with them.

As I dig deeper under into the soil, I see no sign of the black beetle

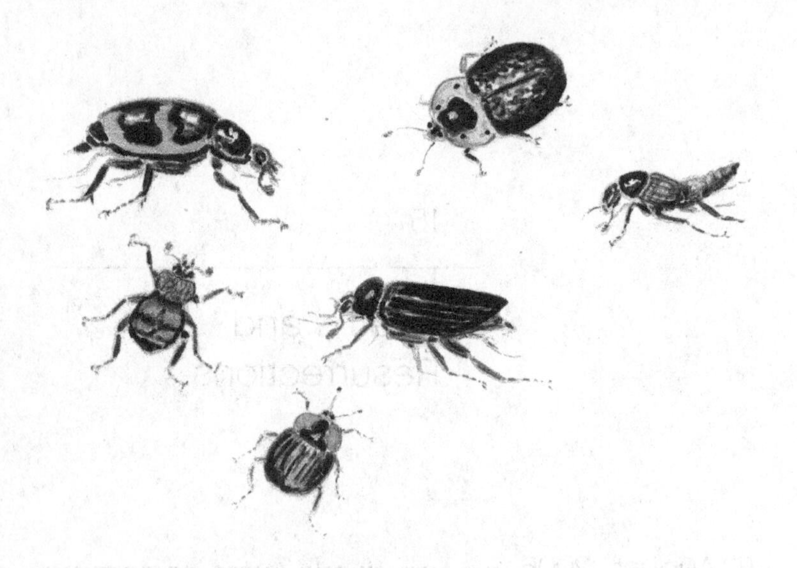

Fig. 29. **Some of the beetles found at the turkey carcass.**

horde. But along the way I discover two species of sylphids. These are round, flattened black beetles with rough upper surfaces; one species has a thorax edged in yellow, and the other is edged with orange. Less numerous but also prominent are two species of staphylinids, or rove beetles. These lithe, elongated animals with tonglike pincers don't look like beetles, because their elytra (wing covers) cover only a small portion of their backs. Their wings are folded up into a small package and tucked underneath those small elytra. One of these staphylinids is black; the other is brown and dotted with shiny gold-yellow flecks. In flight, they sometimes resemble wasps.

When I came back to the turkey carcass twenty days later, the meat was all picked off and dermestid beetles had come and taken their share of the drying remains of skin and bones. No more beetles were visible, but down in the soil I unearthed a gem—a beautiful, iridescent, shiny purple dung beetle that I had never seen before.

EVERY SUMMER THIS HILL BECOMES THE BIRTHPLACE OF countless mammals that range in size from pennyweight pygmy shrews to moose. It is therefore also, necessarily, the dying place of, on average, the same kinds and the same numbers of animals. Most of the small mammals and birds are quickly buried, each by a pair of sexton beetles. This is summer work. The big animals die mostly in the winter, and they have other agents that bring them to the beyond. I think of an old bull moose. His fresh tracks showed that he had staggered through my pine grove; then he collapsed in the snow by a stone wall almost within sight of the cabin.

The moose first became food for coyotes that tore him open, and then several dozen ravens feasted. After the snow melted in the spring there were still plenty of pickings left for the beetles and the flies. Within a month, though, I saw only a pile of hair and bones. Chickadees, and undoubtedly other birds, had come to gather hair for their nest linings, and over the next few years the bones were gradually chewed up by porcupines, squirrels, and mice. There was no waste.

Recently I got a letter from a friend, a former student in California. He wrote:

> *Yo, Bernd—*
>
> I've been diagnosed with a severe illness and am trying to get my final disposition arranged in case I drop sooner than I hoped. I want an Abbey burial. A green burial—not any burial at all—because burial is an alien approach to death.
>
> Like any good ecologist, I regard death as changing into other kinds of life. Death is, among other things, also a wild celebration of renewal, with our substance hosting the party. In the wild, animals lie where they die, thus placing them in the scavenger loop. The upshot is that the highly concentrated animal nutrients get spread over the land, by the exodus of flies, beetles, etc. Burial, on the other hand, seals you in a hole. To deprive the natural world of human nutrient, given a population of 6.5 billion, is to starve the Earth, which is the consequence of casket burial, an interment. Cremation is not an option, given the buildup of

greenhouse gases, and considering the amount of fuel it takes for the three-hour process of burning a body. Anyhow, the upshot is, one of the options is burial on private property. You can probably see this coming. . . . What are your thoughts on having an old friend as a permanent resident at the camp? I feel great at the moment, never better in my life in fact. But it's always later than you think.

—Bill

As far as I was concerned, his feelings exemplify the real and only true religion that I can, in good conscience, honor. So I replied:

I read you loud and clear, old friend. And how amazing to hear your thoughts, when I was just thinking and writing about such things, prompted by a dead turkey that I found in the woods by my field. I was observing the burial/recycling of it by many different beautiful beetles. I was moved to sketch them all, to make it more real, although I'd pass on the art and stick to the ecology when it's my turn, when I'd want no less for myself. I never thought about the cremation part—using up fossil fuels. Thanks for a reminder on that bit. One must live morally toward the Earth, the Creation—burning on a wood pyre is unfortunately not practical anymore. A casket would be for you, as it was for Edward Abbey, our hero, an unacceptable cage for our otherwise free and ever-recycling molecules that would soon become incorporated into Earth's ecosystems.

I'd also not want a spectacle, except possibly by those who sang the Maine Stein Song, even if they sounded like they were trying to raise the dead.

I think I also told him that the practical aspects of his wish are daunting, mainly because overpopulation compromises all our freedoms, from birth to grave. It had not been so in the past. Other friends, perhaps even humans, are already permanent residents here. I once found a piece of worked flint on a little knoll by the brook next to our swimming hole near the beaver lodge. It was revealed on bare earth scratched by my recent logging. It had been deposited during a time when, unlike now, we took it for granted that we were a part of nature. They were of the caribou and the bear. What are we part of now?

No hunter ever had a quarrel with a deer so as to deprive deer of forest, or ducks of marshes. The perceived gulf of separation of "us" from "them" resulted in spiritual isolation from our ecology and our birthright, and it happened only in recent times, at almost the last moment of our existence. It resulted from agriculture, fences, and now also technology that threatens the very last thread of connection. We fence ourselves off from nature. We draw lines and make boundaries. Rather than harvesting our meat from vast herds of bison on the prairies, we destroy the prairies to raise cows and chicks in pens for slaughter only. We destroy prairies and lowland forests and every living thing in them to grow corn and sugarcane to fuel our cars. And then we think we make amends by declaring prairies and lowland tropical forests sacred, and build a fence around a patch here and another one there. We plant trees in rows to make sterile plantations, on the grounds that we should not harvest trees from forests—with the result that forests become fewer as plantations and farms become more numerous. The coffin is a last attempt to place a boundary between ourselves and nature.

Morality, it seems to me, concerns not only doing unto others, but also *being* unto others. I try to connect through a deer in the fall, and perhaps some fish and berries in the summer; and, yes, I do harvest trees and maintain a beautiful forest. The one and perhaps only true religion that I can in good conscience honor is one that encompasses the Earth we walk on and that promotes our well-being and our physical connection to it. Such a religion is based on reverence and respect for maintaining the Creation—whatever origin one either knows or wants to believe. As my friend from California concluded: "Offering oneself to the ravens when the time comes is to me religion at its best."

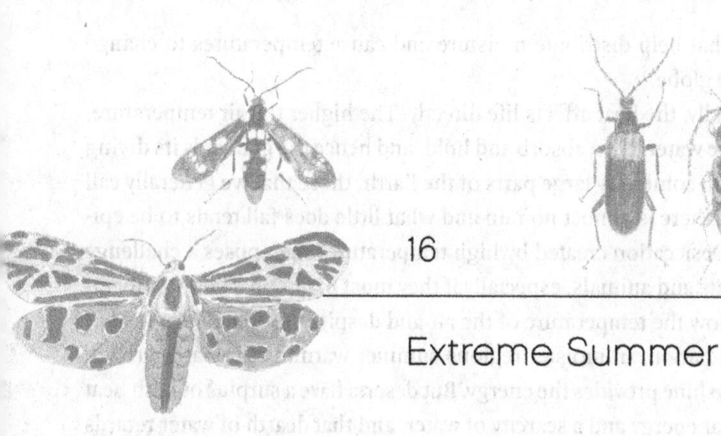
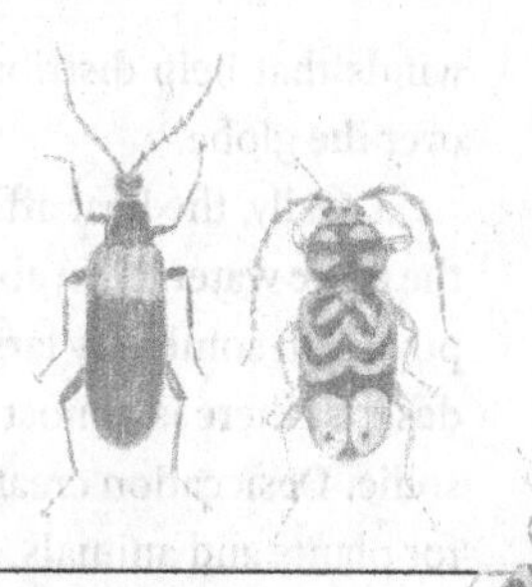

# 16

# Extreme Summer

19 May 2006. APPLE BLOSSOMS WERE IN FULL GLORY today, and ruby-throated hummingbirds, bumblebees, and northern orioles were visiting the flowers for nectar. Blackflies were in full glory as well, when the sun was out. But the sunshine didn't last long, and shortly after dark a thunderstorm blew in. It took only half an hour to pass, but it brought some fireworks. Lightning bolts crackled and ripped across the sky, lighting up the night to make a second of day. Booming crashes followed. One right after the other, they made the earth vibrate and shook the house. Then, after a slight pause, torrential rain gushed down through the trees and pounded the roof. I cannot imagine how birds live through this. How do the baby grackles in the bog keep warm?

SUMMER IS A TIME OF LIFE AND DEATH, IN AN ORCHESTRA of organisms interacting with each other. But summer is determined by two key external variables: temperature and moisture. One affects the other. Thunderstorms come from often distant areas where heat caused evaporation and built up clouds. Rain occurs as the clouds encounter temperatures below the dew point to cause condensation, and the change of the water from a gas to a liquid causes a reduction of volume of the air, which reduces air pressure. The air pressure gradients produce winds that help distribute moisture and cause temperatures to change over the globe.

Locally, the heat affects life directly. The higher the air temperature, the more water it can absorb and hold, and hence the greater is its drying power. In some very large parts of the Earth, those that we generally call deserts, there is almost no rain and what little does fall tends to be episodic. Desiccation created by high temperatures then poses a challenge for plants and animals, especially if they must maintain a body temperature below the temperature of the air and despite the added heat load of solar radiation. In moister regions summer warmth stimulates growth and sunshine provides the energy. But deserts have a surplus of both heat and solar energy and a scarcity of water, and that dearth of water retards or prevents the conversion of the plentiful energy from the sun into the chemical energy useful for life.

Life in deserts confronts hard-edged limits, though often in a context of intense beauty. Life exists there only because of intricate behavioral and physiological adaptations. Field trips into the Mojave and Anza Borrego deserts of southern California opened my eyes to this environment and its exotic animals, which I saw through the lens of work done with George Bartholomew in our lab at my graduate school alma mater, UCLA. "Bart" in turn led me to the research and writings of Knut and Bodil Schmidt-Nielsen, Ray Cowles, and eventually many others who came later. I would have little to say here without their revelations.

Few give a better account of deserts, especially those of southern California, than the pioneer of desert studies, the naturalist Raymond "Doc" Cowles. Cowles grew up in Zululand in South Africa, came to California in 1916 at age twenty, and eventually taught at UCLA. He became an expert in reptilian thermoregulation, and was an academic grandfather of many graduate students and professors who carried on the tradition.

Cowles proposed that the serenity and the severity of deserts makes people who are immersed in the loneliness of these regions into thinkers. In his own holistic views of nature and human ecology he speculated on the meaning of wilderness to society, and he lamented the experiences we were losing. He left for his friends a typed card signed on 1 November 1971. It read: "Raymond Bridgman Cowles, December 1, 1896, at Adams Mission Station, Natal, South Africa, has completed his tour of duty on [he here left a blank] and will now participate in the universal and unending recycling game. This gives notice that his name should now be removed from [reprint] mailing lists." One of his daughters later inserted the date of his death: 7 December 1975.

Ray Cowles, in many ways my own academic grandfather, had experienced half a century of desert field trips on his feet and in his head before he wrote *Desert Journal* (published posthumously in 1977). In it he reminisced about "innumerable campfires and their evening sacrifice of incense from smoldering wood." He was then "sadly reminded that such luxuries, such reverence for the gods of the open skies, are no longer ecologically excusable," and said that "from now on the careful naturalist and his students must be content to enjoy fellowship and worship nature around a noisy hissing gasoline stove for as long as that store of onetime solar energy remains." He anticipated the same "reverence for the gods" that is, as my friend from California recently reminded me, probably no longer possible even in the Maine woods.

Cowles's love for the desert campfire of crackling and smoldering wood and his enjoyment of desert life are revealed in the following passage from a chapter in *Desert Journal*, titled "Around the Campfire":

> Summer or winter, there is something special about sundown and the coming night, and my desert camps were no exception. Not the least was the cessation of work, return to camp, and, in those days of fewer people, the gathering of scanty firewood. I often used cactus skeletons and the roots and stems of stunted shrubs. Soon my camp was rich with fragrance. Food cooked in the aromatic smoke from desert wood has a tang in this clear, unpolluted air unknown outside the arid world. Long before the summer sun has set, the first flight of bats commences, most often the little canyon, or pipistrelle, bat with pale silvery body, black wings and ears. Along the Colorado River they flicker across the sky, bent primarily on reaching water where they replenish the moisture lost during the day, even in their relatively cool rock-crevice retreats.
>
> In the same locale nighthawks by the hundred appear soon after the heat begins to abate. They flutter and sail toward the river for the first drink of the day. During May and June, when many are still incubating or hovering over their eggs to protect

them from the sun's increasing heat, this first intake of water precedes feeding. The birds nest, or more accurately, lay their eggs, on the exposed ground. Throughout the day the relentless sun beats down. Air and ground temperatures may exceed 120°F for hours on end; the direct heat of the sun contributes to what for most creatures would be unendurable conditions. Insulated against heat by its feathers, each nighthawk sits in a self-made patch of shade and comfort. Plumage is as effective in shielding the skin and blood vessels from high temperatures as it is in containing body heat during cold weather.

From time to time they open their enormous mouths and flutter their gular pouch, evaporating some of their small supply of water to keep blood and body temperatures below damaging or lethal levels. But water is so scarce and the day so long that excessively prolonged cooling by this means would dehydrate the birds. I know of no other animals, however, not even the supposedly sun-tolerant lizards, that possess no feathery insulation, that can remain in direct sunlight for so long a time. Many of the smaller lizards will die in minutes under such conditions. Yet the nighthawks, warm-blooded, heat-generating birds, complete their incubation period and care for the young in the unrelenting heat of the desert until all can take flight to the surrounding environment.

In these three paragraphs alone, Ray Cowles eloquently and presciently summarizes volumes of research that came after him. I can add here only a few details to expand on the theme: that birds are preadapted for getting by on less water than mammals because they excrete their nitrogen wastes in a white uric acid paste and thus do not need large amounts of water to flush them out, and they also save water otherwise required for evaporative cooling because their body temperature is 2°F to 4°F higher than ours. Keeping the eggs cool has gone one step farther in the Egyptian plover, *Pluvianus aegyptianus*, which brings water back to the eggs and dampens them to cool them. Similarly, sand grouse in Africa have special feathers on the breast that soak up water so that it can easily be carried back to the nest. The young sip the water from the tips of the feathers, like baby mammals suckling on their mother's teats.

As Cowles indicated, possibly the most effective way for animals to reduce heat input and save precious water is by behavioral adjustments. Desert birds are active mainly in the early morning and evening and take a long siesta in the middle of the day, although some of the larger birds—such as ravens, vultures, and hawks—may soar high in the air, where temperatures are lower than they are close to the ground. It is cooler at night, and most rodents, many reptiles, and many insects escape the heat by becoming nocturnal and staying in cool burrows during the heat of the day. Rodents that are generally diurnal, such as ground squirrels, are heated up temporarily when they venture to run across hot sand, but they then hurry back into their burrow to press their belly against the cool ground and unload their heat.

Avoiding the heat by becoming nocturnal also helps to alleviate the water shortage. The relative humidity is high within a burrow, and so the air cannot suck up moisture from the skin, or from the lungs through breathing. Death in the desert is seldom directly from heat. It comes from dehydration resulting from trying to keep cool. Australian Aborigines have adopted some of the same survival tricks used by other animals. On long walkabouts through hot country, they try to restrict travel to nighttime, and during the day they may bury themselves in the sand to keep from sweating and dying of thirst. Elizabeth Marshall Thomas relates similar strategies, which she learned about from her experiences with the Kalahari Bushmen during the hot dry season. The Bushmen go out early in the morning to hunt for perennial plants whose leaves die to reduce water loss and whose underground tubers are adapted to store water. After a tuber is located by the remnants of its dry vine on the ground, it is dug up and the pulp is scraped out of it and then squeezed to get water to drink. The people survive the heat and dryness of the day by burying themselves in pits dug in the shade. These pits are lined with the tuber shavings, which are then resoaked, but with urine, so that the evaporating water will not be from the precious body stores. At dusk, when temperatures drop, the Bushmen again venture forth to search for more tubers (Thomas 1958, p. 103).

We can tolerate very high air (though not body) temperatures, as was demonstrated (Schmidt-Nielsen 1964, p. 3) more than 225 years ago when Dr. Blagden, then secretary of the Royal Society of London, and some friends, a dog, and some steak spent some time in a room heated to 260°F (48°F above the boiling point of water at sea level). They remained there for forty-five minutes, at which time the steak was cooked but the men and the dog were unharmed (their feet had been protected from touching the floor). Had the air been saturated with water, there would have been no evaporative cooling and it can be confidently said that they would have been cooked along with the steak.

We are not deterred by heat so much as by lack of water. In his book *The Hunters or the Hunted*, C. K. Brain notes that in southwestern Africa, all the Hottentot villages in the Namib Desert are scattered directly along the Kuiseb River. Here the people have dug wells from which they get their water when the river runs dry. Birds there get water from eating insects, and most insects get water from live plants. But one group of Namib beetles of the family Tenebrionidae are an exception. Some of them stay in water balance even when eating only dried plant detritus that blows around in the wind.

These beetles are ground-dwelling, usually large, and black (the melanin absorbs heat but is necessary to protect them from damage by ultraviolet light). They live on the sand surface. Those that live on the hottest sands have stiltlike legs to reduce heat input from below. Others

Fig. 30. A Namib Desert tenebrionid beetle, which elevates itself above the most intense heat at the ground surface.

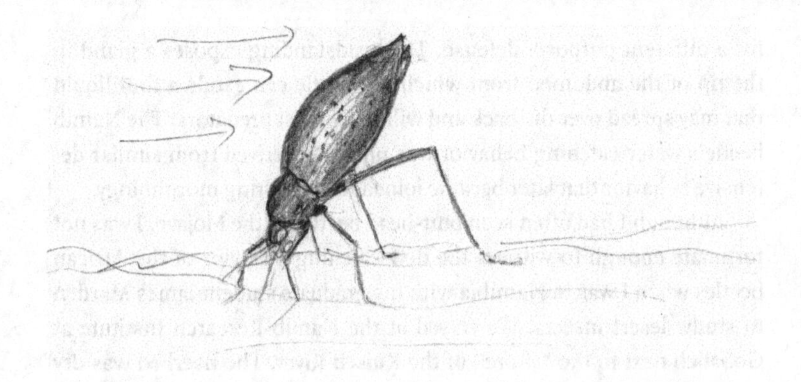

Fig. 31. **A Namib Desert tenebrionid beetle that catches water from moist air blowing in from the Skeleton Coast by doing headstands. The water condenses on its back in tiny droplets, which then coalesce and run down to its mouth.**

reduce heat input from above, from the sun, by light-colored wax on their black backs. But even then there is still the problem of getting sufficient water, and there is no standing water and no rain when they are active. Although they are subjected to a desiccating environment during the daytime, at night temperatures in the Namib typically drop and wind from the Atlantic coast may sweep in with moisture-laden air. The beetles then orient themselves by standing on the sand dunes with the head straight down and the abdomen up into the air. Water condenses on the beetle's back and flows down in droplets to its mouth.

The beetle's amazing behavior is cobbled together by evolution from structures and behaviors that previously had other functions. Their backs are modified wing covers (elytra) that no longer cover any wings but serve instead as physical protection for the body. But in these beetles the elytra have taken on an additional, very different, and novel function. All tenebrionid elytra are sculptured in various patterns. In these beetles they have a pattern of bumps that helps capture vapor molecules into tiny droplets. Waxy valleys between the bumps channel the water droplets so that they coalesce and roll down into the mouth. I recalled seeing similar tenebrionid beetles in our southwestern Mojave Desert, where they are sometimes derisively called "butt-head" beetles because here also they stand with their butt in the air. But in this case they do their headstands for a different purpose: defense. The headstanding exposes a gland in the tip of the abdomen from which the beetle can exude a foul liquid that may spread over the back and will repel most predators. The Namib beetle's water-catching behavior was probably derived from similar defensive behavior that later became joined to an existing morphology.

Although I had often seen butt-head beetles in the Mojave, I was not fortunate enough to witness the dew-catching behavior of the African beetles when I was in Namibia with my graduate student James Marden to study desert insects. We stayed at the Namib Research Institute at Gobabeb next to the "shore" of the Kuiseb River. The riverbed was dry at the time, but it was the only place where we saw trees and shade. The trees' roots tapped the groundwater, and that water fed insect fauna. We saw innumerable black tenebrionid beetles running in haste. They were mostly racing in pairs, with the female always in the lead. Jim made this curious phenomenon the focus of his study.

We saw no sign of free water. Yet during World War II two German geologists from nearby Windhoek—Henno Martin and Hermann Korn, with their dog, Otto—managed to hide out here undetected for two and a half years (to avoid being put in an internment camp). They lived like Robinson Crusoe during those years, and Martin later wrote a book about their experiences. In it, he describes the effect that water has on life in the desert. Martin and Korn had experienced a drought in the Namib for several years, and one night they heard thunder:

> I had never before in my life heard [such thunder], or experienced such a cloudburst—and now the scorched and battered life began to raise its head again—within hours, bushes that had looked dead began to show shoots of green and in the shade of rocks ferns began to unroll delicate light green leaves. The desert was alive everywhere: seeds that had lain dormant for years came to life and pierced the crust of the earth; almost overnight the balsam bushes covered themselves with green leaves like young birches; the first yellow flowers opened their petals to the sun, and once again we found the speckled eggs of the quail amidst the grass and stones; and the lukewarm water on the pools swarmed with little crab-like insects whose eggs had survived the years of dryness and scorching sunshine.

Adaptations of plants to deserts include dormancy and a variety of structural and behavioral adaptations. The majority of desert plants depend on a strategy that capitalizes on small size. They are annuals that spring up from dry, dormant, heat-resistant seeds. Some of these seeds may wait up to half a century before they are activated. The plants' challenge is to be quick enough to respond to rain so that they can produce their seeds before the earth dries up again, while not jumping the gun to start growth until there is sufficient water for them to grow to maturity for seed production. Some achieve this balance on a tightrope by "measuring" rainfall. They have chemicals in their seeds that inhibit germination, and a minimum amount of rain is required before these are leached out. Others have seed coats that must be mechanically scarred to permit sufficient wetting for germination, and the scarring happens only when they are subjected to flash floods in the riverbeds where they grow. A plant in the Negev Desert releases its seed from a tough capsule only under the influence of water through a mechanism that resembles a Roman ballistic machine. Its two outer sepals generate sideways tension that can fling two seeds out of the fruit, but the two seeds are held inside by a lock mechanism at the top. However, when the sepals are sufficiently wetted, then the tension increases to such an extent that the lock mechanism snaps, and the capsule "explodes" and releases the seeds (Evenari et al. 1982, p. 399).

In moist regions where it rains predictably (though not necessarily in abundance), we help agricultural plants to capture the precipitation by scarring the soil to facilitate the infiltration of the water into it, and hence into the roots. Least runoff and maximum water absorption are achieved by plowing the soil. However, such a strategy would not work in a true desert such as the Negev. A different program is required there because rain is infrequent and plowing would facilitate only the evaporation of scarce water from the soil. The solution applied by the peoples who inhabited the Negev in past centuries was a practice they called "runoff farming." Farmers had mastered harnessing the flash floods that rush down into the gullies by catching the runoffs—not only by making terraces but also by building large cisterns into which the water was directed to be held for later use. Remnants of these constructions still exist.

Water-storage mechanisms have been invented by other organisms living in deserts, but mainly through modifications of body plan. Many plants, especially cacti and euphorbia, have the ability to swell their roots or stems with water stores. Possibly the most familiar is the saguaro cactus, *Carnegiea gigantea*, of the Sonoran desert in the American southwest. It has a shallow root system that extends in all directions to distances of about its height, fifty feet. In one rainstorm the root system can soak up 200 gallons of water, which are transferred into its tall trunk. This trunk is pleated like an accordion and can swell to store tons of water that can last the plant for a year. The cactus has no leaves, but the stem is green and can photosynthesize and produce nutrients as well as store water. The saguaro's survival strategy requires it to grow extremely slowly. But it lives a century or more.

Some desert animals similarly store water. The frog *Cyclorana platycephala*, from the northern Australian desert, fills up and greatly expands its urinary bladder to use as a water bag before burying itself in the soil, where it spends most of the year waiting for the next rain. While in the ground it sloughs off skin and forms around itself a nearly waterproof cocoon that resembles a plastic bag and reduces evaporative water loss.

Desert ants of a variety of species (of at least seven different genera) in American as well as Australian deserts collectively called "honeypot ants" have evolved a solution that combines water storage with energy storage. Ants typically feed each other; and some of the larger worker ants may take up more liquid than the others, and others may bring more. Those that take the fluid may gorge themselves until they distend their abdomens up to the size of a grape, by which time they are unable to move from the spot. They then hang in groups of dozens to hundreds from the ceiling of a chamber in the ant nest, where they are then the specialized so-called repletes that later regurgitate fluid when the colony members are no longer bringing the fluid in but rather needing it. In western North America about twenty-eight species of one genus, *Myrmecocystus*, have adopted the storage strategy of water and sugary secretions that are secured from aphids, flower nectar, and other plant secretions when summer is not yet too severe.

Animals' solutions to the extremes of desert summer have also been exploited by people. In the Australian deserts the Aborigines have learned to find and access the water-holding frogs and use this resource as a last resort in times of need. In central Australia also the repletes of one honeypot ant species, *Camponotus inflatus*, are large enough to be

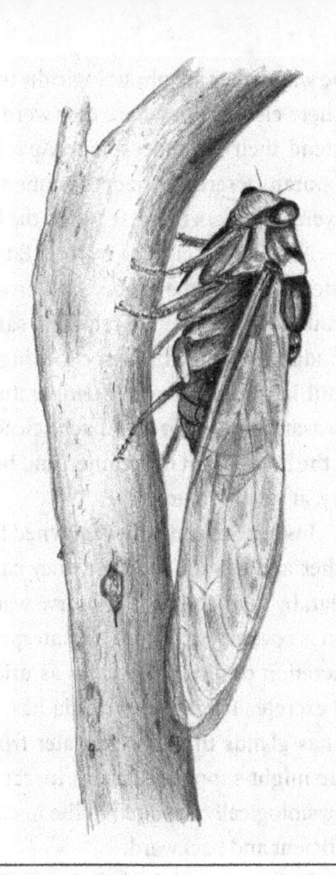

Fig. 32. The Apache cicada is active during the hottest part of the day in the summer, when most animals try to escape the heat.

commonly used by Aboriginal peoples. Those of *Myrmecocystus mexicanus* in the southwest in North America, who store water or honey or both, were also used by native peoples (Conway 2008). The Bushmen of the Kalahari Desert, instead of exploiting the fluid stores of frogs and ants, use the shells of ostrich eggs as containers for underground water storage caches; but as mentioned previously, when they exhaust these stores they resort to water stored by plants in underground tubers.

For those who solve the water problem, the desert can be a haven. For peoples living in the American southwest, the Namib, and the Negev, the desert has often been a refuge from persecution. Under what circumstances except necessity would people be so ingenious and hardworking as to try to make the desert bloom and grow crops? Why would animals live where they are physiologically tested to the limits of their endurance? Where else except where they were severely tested would they evolve to extend their tolerances? The Apache cicada, *Diceroprocta apache*, of the Sonoran Desert of Arizona, is one such animal. It not only tolerates the severe summers there; it courts the heat.

As with the cicadas in New England, and those at numerous other places all over the globe, the larvae of the Apache cicada live underground, where they are relatively safe, and they could potentially emerge as adults at a time of their choosing. Here in New England, cicadas wait until late summer when temperatures are, by our standards, benignly pleasant. Not so for the Apache cicadas of Tucson, Arizona. They emerge in the hottest part of summer, and become active at the hottest part of the day, at nearly 110 to 116°F.

Insects are generally renowned for their ability to retain water. Where other animals die of thirst, they can stay hydrated, mostly by avoiding heat, by their ability to conserve water because their watertight exoskeleton is covered with layers of waterproofing lipids and waxes, and by their excretion of nitrogen wastes as uric acid that requires negligible water to excrete. The Apache cicada has, literally, prominent holes in it, and it has glands that excrete water from these holes. Not knowing more, one might suppose that this insect is active at the wrong time and also physiologically unsuited to life in extreme summer. Its design seems inefficient and backward.

It took two biologists, James E. Heath from the University of Illinois and Eric C. Toolson from Arizona State University, to unravel the cicada story, which is one of exquisitely elegant desert adaptation. Heath deduced from his studies that the cicada's seemingly anomalous active time is the time when potential predators—both birds and wasps—have fled the field because they can't stand the heat. Toolson found out that cicadas can stand high temperatures because they have glands that function like sweat glands to provide evaporative cooling in emergencies, as when the males are exerting themselves by calling deafeningly to attract females for sex.

The cicadas' ability to defy the summer extremes, and thereby to escape enemies, would not be possible without a constant, reliable source of water. And as members of Homoptera—the aphids and their relatives—cicadas are preadapted to get that water. Near Tucson in the summer, the Apache cicadas perch all day in the shade of a paloverde branch in an arroyo and tap water from deep down in the soil. The means of access to that water are the deep roots of the bush, which grow as much as sixty feet down to the water level. The water is piped to the twigs that the cicadas tap into with their sucking mouthparts.

Thermal wars are also waged directly, in combat between one insect and another. The Asian honeybee *Apis cerana japonica* faces a serious predator, the giant hornet *Vespa mandarinia japonica*. Hornet scouts invade beehives and if successful recruit their nest mates to come in force and devastate a beehive. The hornet is far too big and heavily armored to be killed forcibly by the much smaller bees. However, these honeybees have evolved a strategy that compensates for their size deficit. They pin a hornet down by clustering in hundreds around it to form a ball, and then they shiver and produce enough heat to raise the temperature at the center of the ball, where the hornet is, to 118°F. That temperature kills the hornet but is still a degree or two below the upper tolerance of the bees (Ono et al. 1995).

A slightly different story is played out by thermal warriors during the summer close to my home in Vermont and the woods in Maine. In this case the white-faced hornets, *Dolichovespula maculate*, whose summer colony strategies I discussed previously, are the beneficiaries of the thermal strategy. It is often hot in the daytime during our summers, but in early and late summer nighttime temperatures commonly dip to 38°F or lower, and they may remain there early in the morning. Such temperatures are so low that many small insects hunted by these hornets cannot fly. The hornets hunt by cruising over the foliage and pouncing on any contrasting object that may be a fly or some other unsuspecting insect. They have their best chance of success when their prey has slower reaction times or cannot fly off, or both; and that is in the early morning, when it is still cool. This is when the hornets, with a muscle temperature near that of our own, leave their warm insulated paper nests in force to hunt. They are larger than their prey, and have the additional advantage of flying at low temperatures because their exercise, both flying and shivering to prepare to fly, results in more heat retention than is possible for the much smaller prey.

In my opinion the most extreme and the most beautifully elucidated thermal warrior strategy is one found during the extreme summer in the

Sahara Desert. This is the story of the silver or "fast" ants, genus *Cataglyphis*, as unraveled by Rüdiger and Sibylle Wehner and colleagues from the University of Zurich, Switzerland. These ants are remarkable because they preferentially forage at midday, when ground surface temperatures reach 145°F. They tolerate a very high body temperature of 129°F, but because of their small size they would reach a lethal temperature within seconds after coming out of their subterranean nests and stepping onto the sand. They are much too small to cool off by the evaporation of water; instead, they survive by pausing frequently to cool off by climbing dry stalks that serve as thermal refuges. The question is: why don't they go out at night like most other desert dwellers, when they would not desiccate so easily and would automatically escape the danger of being killed by heat?

The Wehners discovered the answer in the ant's hunting strategy. These ants are fast, but not fast enough to run down live prey. They specialize in insects that have been incapacitated or killed by heat. The hitch is, however, that they cannot afford to leave their safe nests until the sand is hot enough to prevent their own major predators, lizards, from being out. As a result of this thermal tightrope, these ants must wait at their nest entrances, not risking a mass exodus, until it is hot enough for the lizards to have retired from the field but not so hot that the ants themselves would be incapacitated.

A focus of study for the Wehner group has been trying to understand the ant's homing ability. The ants forage for insect prey incapacitated or killed by heat. This requires extensive searching, and taking many twists and turns in their paths. Moreover, at the end of a foraging trip—after finding prey or after temperatures rise to excessive levels—the ant has to find its way back. And it may have to get home in a hurry, since sand temperatures are often very high and the ants can stand only so much heat.

The ants' safety margin, with regard to their physiological tolerances, depends on a combination of rapid running and unerring homing ability. The Wehner group determined that the ants' homing ability is the result of an amazing mental feat. The ants compute where they are at any one time by integrating the turns and distances of their travel ("path integration"), and then use the sun's location from the pattern of polarized light in the sky as a compass to determine the homeward direction

Fig. 33. The long-legged "fast ant," *Cataglyphis*, of the Sahara Desert comes out of its relatively cool underground nest to go foraging on the sand surface in the daytime at near its thermal death point. It reduces direct solar radiation by being pseudo-erect, elevating its abdomen.

and distance. Near the end of their run by the nest entrance, they also use landmarks, if any are available.

The ants' lives outside their subterranean nests are hazardous, and they venture out only near the end of their lives. An analogous situation among humans would be drafting soldiers in old age because they have already lived and contributed to society, rather than sending youths into danger who have still much to do and give.

Various anthropologists and physiologists have remarked that we, too, are creatures that may have begun in an environment of extreme summer. We combine the extreme sweating response of the Apache cicada with the extraordinary running, hunting, and homing abilities of the *Cataglyphis* ants of North Africa and Asia (other genera of ants in the South African and Australian deserts have a similar lifestyle). In combination with imagination, they have given our distant ancestors (as well as some contemporary tribes) a thermal advantage to avoid being eaten by predators and to run down antelope.

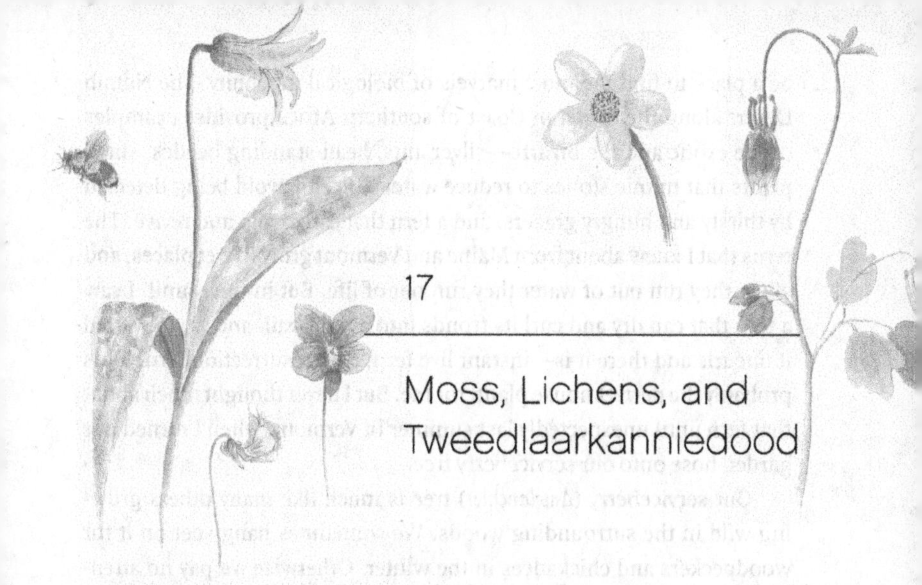

# 17

# Moss, Lichens, and Tweedlaarkanniedood

THE ROBINS AND THE PHOEBES ON OR NEXT TO OUR house hopped out of their nests, adult-size, at two weeks of age at the most. They were aided first by the warmth of a brooding parent, and then they warmed themselves by their own metabolism. The plants' growth, stimulated by the warm summer days, was as impressive. Rachel kept track of what grew in the garden, and I was more focused on what went on outside it. On my daily jogs past a beaver pond I was especially impressed by how fast the stump shoots grew where the beavers had chewed off trees. Some ash shoots went up nine feet in a single season, and red maple shoots grew as much as sixty-six inches. They had been growing at a steady rate of almost an inch per day for the whole summer. Surprising as the fast growth was to me, I was even more impressed by how quickly it could come to a halt. Most trees stopped lengthening their twigs entirely by mid-June, when there were still three months of summer to come, but vines and some tree stump shoots (those in direct sunshine) kept right on growing at the same furious pace. Warmth and sunshine may translate into growth, but only if everything else is equal. In deserts there is plenty of both, but growth tends to be very slow.

Deserts are a source of marvels of survival in extreme summer, and so an extreme desert—one with the least water and the most heat—should be a place to find the most marvels of biological ingenuity. The Namib Desert along the Skeleton Coast of southern Africa provides examples of the exotic and the bizarre—silver ants, head-standing beetles, small plants that mimic stones to reduce water loss and avoid being detected by thirsty and hungry grazers, and a fern that can dry up and revive. The ferns that I knew about from Maine and Vermont grow in wet places, and when they run out of water they run out of life. But in the Namib I saw a fern that can dry and curl its fronds into a tight ball, and when wetted it uncurls and there it is—instant live fern, the "resurrection fern." It is probably the perfect house plant, for me. But I never thought much about that fern until unexpectedly last summer in Vermont, when I turned our garden hose onto our serviceberry tree.

Our serviceberry (*Amelanchier*) tree is much like many others growing wild in the surrounding woods. We sometimes hang suet on it for woodpeckers and chickadees in the winter. Otherwise we pay no attention to it—except in May, when, several days before the trees leaf out, it erupts for a week in a mass of white flowers. That is also when the ground thaws, the time people used to bury the dead and hold funeral services here in New England (hence the serviceberry name). By June this tree bears purple berries (hence its other common name, Juneberry). Even before they are ripe, these berries already attract cedar waxwings, and then in late June and early July they also attract robins, along with rose-breasted grosbeaks, purple finches, wood thrushes, catbirds, veeries, tufted titmice, and cardinals.

In 2007 our summer, like most summers, had long dry spells. I got out the garden hose to water our serviceberry tree, remembering all the birds who feed there. I did not want its roots to dry up, because as with most plants, even a temporary absence of water kills. As I was idly spraying the ground beneath this slender tree I noticed for the first time what I had undoubtedly seen hundreds of times before: yellowish green moss on the rocks under the tree. Surely this moss would be totally dry! I bent down and peeled off a patch of it—dry indeed, dryer than a bone. I realized then that this moss was already many years old, and it must have been dry on many occasions during past summers.

I peeled off a handful of moss and put it into a bowl of water in the sun, and—presto—it soaked up water like a sponge; in seconds its slender fronds expanded and became vibrant green. It was just like the resurrection fern in the Namib Desert, which I had thought to be a unique marvel. In an hour small silvery bubbles (oxygen) formed on the submerged moss—it was respiring; it was alive. I put a sample of the moss back on the rock, where it again dried, more quickly than the clothes we put out on our clothesline. I then collected samples of five other species of moss from our woods. I dried and then wetted them at intervals of months. They froze outside in the winter, and then I brought them in again to dry out. I let a portion of them stay almost totally dry for six months, and when I dunked them in water they again soaked up water in seconds, and then looked as fresh and green as when I had picked them. I did the same with three species of club mosses (*Lycopodium digitatum*, *L. clavatum*, and *L. obscurum*). They all died when they dried, and once they dried, it was difficult to get them to absorb water. For a further contrast with real moss, I picked eight kinds of green herbs and let them dry. After a week the dried leaves looked a dull green, but when wetted (with difficulty) they were black and dead.

In mid-November, while I was up in a balsam fir tree in Maine, sometimes for hours at a time, I had further opportunity to admire the miracle of mosses. Right next to me on the limb where I was perched, I counted at least three species each of mosses, growing in part intermingled with as many species of lichens. Every limb was loaded with both, as were those of the neighboring trees. The ground beneath me was covered with already browning fallen leaves, but the rocks that poked out among them were covered with vibrant, luminous green cushions of moss in moist areas, and with lichens in more dry and exposed areas.

The lichens from the branches dried as quickly as moss. They also absorbed water as quickly, and then they seemed as vibrant as any in spring and fall, when they are normally wet all the time. The water-absorbing property of moss is of course well known, and sphagnum moss, especially, is a traditional diaper material used by northern peoples. In their deathlike state lichens are protected by several antibiotic chemicals, should any microbes attempt to consume them. Lichens are a cooperative association of a fungus and an alga, in which the alga provides carbohydrate for the fungus, and the fungus provides minerals and shelter for the alga. The summer had revealed common marvels, which I had seen before but not noticed. They remind me of the resurrection fern of the Namib, but another unique plant from there, the two-leafed *Welwitschia mirabilis*, is in a category by itself.

*WELWITSCHIA* IS NAMED AFTER FREDERICK MARTIN JOSEPH Welwitsch, an Austrian medic, naturalist, and collector who first found it in Angola on 3 September 1859, the year that Charles Darwin published *On the Origin of Species by Means of Natural Selection*. This plant resembles no others, and its evolutionary origin is still an enigma. It is the sole representative of its genus and the sole species within its plant family. Its Latin species name, *mirabilis*, means unique or wonderful. The Afrikaners of South Africa also call it Tweedlaarkanniedood (literally, "two-leaved-cannot-die"). This plant is also radically different physiologically from all other desert plants; its two huge leaves stay green and hydrated continuously, and it may live more than 1,000 and possibly 2,000 years.

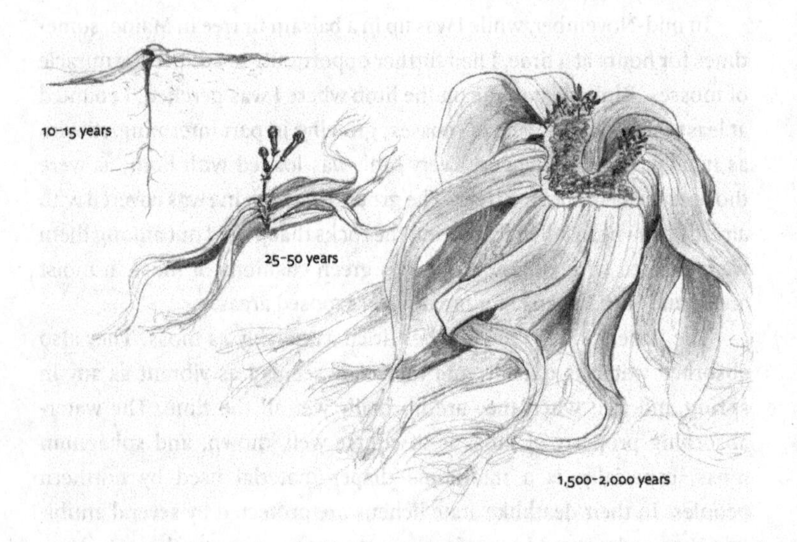


Fig. 34. *Welwitschia* plant, a unique denizen of the Namib Desert that does not shed its leaves as most other plants do, and that stays hydrated when others dry out. It has two lifelong leaves that may grow (as they fray) for more than 1,000 years.

Other plants adapt to extreme heat and drought by having no leaves, or small leaves that are shed when water becomes scarce. *Welwitschia mirabilis* has two straplike leaves that are more than a yard in diameter and several yards long, and they are never shed. Like hair, they just keep growing from the base, and gradually wear off or disintegrate at the end. The functional (living) part of the leaf may be up to seventy years old and earns the distinction of being the oldest known living leaf tissue.

Leaves lose water primarily through their stomata, the pores needed for gas exchange, and most desert plants minimize the number of these microscopic openings and locate them on the lower leaf surface. *Welwitschia's* leaves have about 250 stomata per 0.0016 square inch, more than most temperate and tropical plants, and these are located on both the upper and the lower leaf surfaces. In short, this plant is both a botanical and a physiological paradox of desert adaptation that could not have been fully appreciated by Welwitsch when he first found and described it, even though he wrote, "I am convinced that what I have seen is the most beautiful and majestic that tropical South Africa has to offer."

Leaf stomates typically stay open during the day to allow carbon dioxide to diffuse in so that it can become fixed into small carbon compounds during the process of photosynthesis, in which sugar is produced. Water then necessarily evaporates, leaving passively through the open stomates, especially when the leaves are heated by the sun and the air is dry. However, most desert plants (in this case including *Welwitschia*), have evolved the ability to conserve water by closing their stomata during the day when water loss would be high, and they have a special mechanism that still allows them to perform photosynthesis. They conserve water by opening their stomates at night when the air is cooler and more saturated with water. And the carbon dioxide then enters (it can't be used for photosynthesis just then, because there is no sunshine) following the diffusion gradient (from high concentration outside to low concentration inside the leaf). The gas concentration is lowered inside the leaf as gas that enters is removed (by being incorporated and thereby stored) in malic acid. Then, in the daytime, the reaction reverses itself—the malic acid breaks down and releases the carbon dioxide, which stays inside because the stomates are closed then. The stomates are closed to conserve water, and the carbon dioxide then held captive is available to be used in the normal way to make sugar by photosynthesis. Despite this water conservation—a trick that is also used by many other desert plants— *Welwitschia* still needs water, and it employs a mechanism like that of the "butt-head" tenebrionid beetles who inhabit the same environment: the capture of water condensed from the air. It has no mouth, though, to suck this water up.

*Welwitschia*'s stomates are arranged in troughs formed by parallel ridges on its leaves. These ridges are much like the ridges on the tenebrionid beetles' backs, and function as they do in catching water vapor. Water that condenses on cold nights runs between the ridges, along the troughs, where it is absorbed by these stomates.

According to NASA's definition, life is "a self-sustaining chemical system capable of undergoing Darwinian evolution." Almost all signs of the chemistry of life point to one origin here on Earth. Despite the great diversity of forms, the internal machinery of life on our planet may be so conservative because all life here evolved from common ancestry: all organisms are constrained by their evolutionary history. However, life as we know it is also constrained by the physical properties of the elements that compose it, and that possibly further constrain it into its specific configurations by temperature and pressure. We assume that life is not likely ever to be radically different from ours, or in all the billions of solar systems where it is almost certain to exist, to have existed, or to exist in the future. When (in 2007) astronomers discovered Gliese 581c, another of 227 new planets discovered thus far, it generated excitement because its distance from its sun and its size suggested that its temperatures probably ranged from 32° to 104°F, conditions that are just right for the possibility of liquid water. Hence it is one of the first planets thought hospitable to life. We automatically make the assumption that life requires certain conditions regardless of where it is found and that it would be like our life. But who is to say that oceans of ammonia or methane might not evolve other, bizarre, life?

In his book *The Fitness of the Environment*, Lawrence L. Henderson argued (in 1912) that the properties of matter, especially water and carbon, are requisite for life to evolve, and that possible abodes of life not unlike Earth must be a frequent occurrence in space. George Wald (in the preface to a reissue of this book in 1958) also assumed that life must exist elsewhere and it would be "life as we know it, for no other kind, I believe, is possible." Henderson agrees, but concludes, "The biologist may now rightly regard the universe in its very essence as bio-centric." Others now extend this to homo-centric. But if *Welwitschia* could speak it would say, "God has been kind and thoughtful to me above all others. He has given me two leaves, no more, no less, just exactly the right number that I need, and he has made them to last me a lifetime, and he has put me here in this environment that is so hospitable *for me* that I don't need to move from the spot and can exist here. He supplies all my needs so that I can live without worries for centuries. The temperature—extreme summer to anything else—is perfect. I never overheat, and food is provided from the ground and the air. Water and carbon dioxide come to me in the foggy air at night. I'm in paradise. He has foreseen every little thing to make my life complete. Therefore, when he created the world, he must have had me, specifically, in mind." Or the serviceberry tree, the lichens, and the mosses that thrive exactly two feet from my back door.

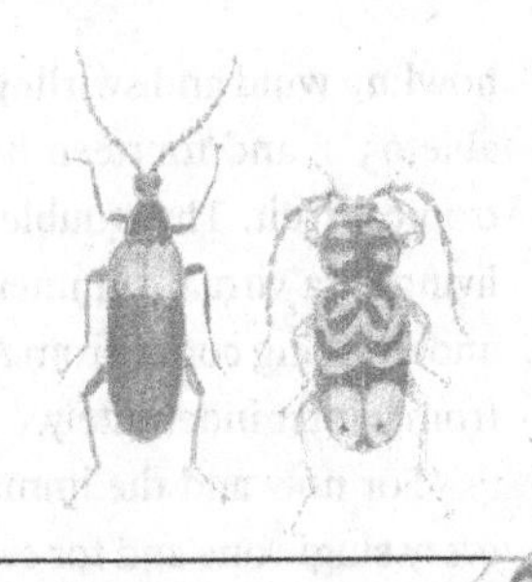

# 18

# Perpetual Summer Species

12 August 2006. THE STUNTED SWAMP MAPLES ARE starting to turn red and are dropping leaves. The crickets' monotonous chirping refrain—before they mate, lay their eggs, and die—is constant. In contrast, the birds are almost silent. But they may be more restless and on the move through new territory—two ovenbirds (common warblers living in deep shady woods) just hit the windows and were killed. I am also already switching modes, but downward. I curl up in bed a little earlier, sleep later, and eat more. The farmers have harvested their second crop of hay. I've finished the woodpile, and we're canning tomatoes and string beans. Signals indicating the end of summer are all around.

TWO VERY DIFFERENT GROUPS OF ANIMALS LIVE IN A PERpetual summer, or nearly perpetual summer, in the far north and deep into the south. The first are birds that migrate from one summer in the north to the other one at the other end of the globe. They can always live in a summer world, thanks to energy-rich berries and heroic sustained exercise. We have come close to imitating them. We manage the same trick of living in perpetual summer, although not by strenuous biannual migrations but by creating and retreating into "climate bubbles." Temperatures outside may be minus 50°F and the outdoors can be dark with howling wind and swirling snow, but we can be experiencing a comfortable 65°F and fourteen hours of light per day while we feast on fresh tropical fruit. The trouble is that a population of hundreds of millions living in a virtual summer while eating bananas from Central America and drinking coffee from Africa probably can't sustain wresting summer from winter indefinitely.

For now and the immediate future, each household that I know of, every single one and for every day throughout six months of winter, consumes vast quantities of fuel imported from thousands of miles away to keep the occupants warm and for cooking, lighting, transportation, and—directly or indirectly—for almost everything we do and own. For six months we can't grow any food. And we insist on building more homes here all the time; almost weekly there are new patches of monstrous new houses that arise like mushrooms up out of the ground in our neighborhood, and each and every one of them requires more and more of the same fossil fuels. Without them vast stretches of the North American continent would virtually overnight be depopulated. And with agrofuels to build and sustain them, vast stretches of the Earth's most beautiful southern ecosystems would have to be sacrificed, creating biological deserts to sustain our northern perpetual summer.

It is at this point tempting to spout personal and political views. But factual scenarios are scarcely either. There is the necessity of maintaining sound natural ecosystems—those that sustain the life of all animals that evolved in them and that live there in a complex unity.

I am an optimist. There is a way. As Thoreau wrote, "Men think it is essential that the Nations have commerce, and export ice, and talk through the telephone, and ride thirty miles an hour." He meant they are mistaken. I believe Thoreau was a happy man. People have lived happily in a small cabin in the woods where they had none of the amenities such as refrigerators, oil furnaces, electric toasters, cars, telephones, television, running water, etc. Some who have experienced it even think with nostalgia of such a presumably deprived existence.

It is unlikely that we will, or even can, change our lifestyles radically enough to make much difference. It is madness to suppose we would make a significant difference by using more energy-efficient lightbulbs and using agrofuels rather than oil, or that city dwellers can or would take up a rural farming or a hunter-gatherer lifestyle: given our numbers, there is no land. There is only one thing to do that will have an almost immediate effect (say, in a century or two): radical reductions of population. Ironically, if we do take that route then we *can* have everything—cars, jetliners, televisions, and all the rest, even perpetual summer. With a low population we could subsist and get by, in perpetuity, with the most efficient method yet devised for capturing solar energy—trees.

We can cut down some of the most beautiful creations imaginable, but out of forests. That requires having more forests rather than creating tree plantations. We need two things: clear vision and also a spiritual imperative so that we will focus on the ultimate ecology, not the proximate economy. The increase in human happiness of *future* generations that this simple solution would create staggers the imagination, and the vast misery that would result if we do not adopt it is almost too horrendous to contemplate. Those are the "knowns." The solution is obvious. The treating of symptoms is opinion and hype.

I ask here instead how we got to where we are now. To start simply, I think we can learn a lot from—yes—hair. If one accepts the almost universally applied premise that we evolved from furred apelike ancestors, then our present insufficient amount of insulating body hair indicates that we evolved while being subjected to more overheating than was experienced by them while other (furred) lines became present-day apes. (An alternative hypothesis, which needs scarce consideration, is that we became naked to shed lice. If that were true, then any number of other primates would also be naked.) That is, not only did we not need insulation; it was a liability in terms of survival. We were, therefore, spawned by a perpetual summer world.

When *Homo sapiens* first spread out of Africa about 150,000 years ago (plus or minus a few tens of thousands of years), we were, as now, already defurred or nearly so. However, by then we were also clever enough to co-opt the fur of other animals who had already adapted to a cold environment. We don't know precisely when that happened, but thanks to lice, DNA technology, and clever sleuthing by the geneticist Mark Stoneking at the Max Planck Institute of Evolutionary Anthropology, it looks as though we became clothed about 115,000 years ago. The remarkable closeness in the two dates—spreading out of Africa and becoming clothed—is probably not coincidental.

Lice are ectoparasites (parasites living on our skin rather than under it), and ectoparasites are remarkably species-specific; each kind of bird or mammal has its very own louse and flea species living on it because each is an island with regard to the others. However, even though any one nonhuman mammal species has the dubious honor of hosting only one of each louse or flea species, humans are unique: we have three species of lice. They are head lice (*Pediculus humanus capitis*); body lice (*Pediculus humanus corporis*), which live primarily in clothing; and pubic lice (*Pthirus pubis*).

Thanks to the logic of evolution and DNA technology, we know that DNA accumulates gene changes, generally at a steady rate. Therefore, by comparing the number of gene changes between two animals, we can use DNA to determine relatedness, and we can use the changes as a "clock" that tells us *when* the divergence occurred. The data indicate that head and body (*Pediculus*) lice had a common ancestor about 114,000 years ago. Logically, these lice would not have diverged to become two species unless they had two different habitats to adapt to. The choice pubic spot had already been taken much earlier by a more distant relative, *Pthirus pubis*. Stoneking deduced that when we were still fully furry—long before we were human—we had only one other louse habitat, the fur covering the rest of our body. We became naked under the heat in the tropical African savanna, and only the head hair (which has special significance, as I will discuss later) remained as the second ecologically suitable louse habitat. When we came out of Africa about 150,000 years ago, we were probably still naked, but we would not have progressed very far north without wearing clothes. The lice took up residence in those clothes, and they needed different behavior to prosper in that new and different habitat, next to our warm body rather than on the head. A population of the original colonists stayed in our head hair, but the clothes-loving lice diverged and perfected to live in their new habitat. As they adapted, their offspring would have been disadvantaged by being saddled with the lifestyle of their old haunts. Similarly, the head lice would also be disadvantaged by inappropriate lifestyles, so isolating mechanisms evolved, and eventually the two lice could no longer interbreed and the line split into the two species.

More interesting, perhaps, is the obvious question: why were we naked in the first place? If we came out of Africa naked or nearly so, and if the apes' and our common ancestor was probably hairy, as all apes still are, then why did we *become* naked? I think the best hypothesis to account for our nakedness is that we derived from a very special ape-man, an endurance predator who depended on rapid and prolonged locomotion in the heat in order to compete with other predators, primarily sprint specialists. We can still compete with cheetahs, lions, and leopards in running down antelope, but we can do it only in the midday heat. And the reason is that we have the mental capacity to pursue a goal that we can neither see nor smell but that we can imagine. Additionally we have a unique suite of adaptations to deal with internally generated body heat under the blazing sun. They include our nakedness, our ability to route blood to the surface of our extremities so that our veins bulge at the surface of exposed skin, and our ability to sweat profusely over the skin. These are capacities needed by hunters who get their edge through endurance in the heat.

A recent review article (Rantala 2007) argues that the cooling hypothesis "does not bear close scrutiny." Perhaps, if one discounts the zoological perspective: that the *predecessors* of *H. sapiens* differed from other hominids in being erect and needing to hunt at noon under the direct overhead sun in order to compete with the large carnivores who rest then. Although feathers and hair on the dorsal surface insulate other desert animals from direct solar radiation, most have "thermal windows"—areas of very thin hair or no hair, as on the bare bellies and flanks of desert antelope, the areas less exposed to the direct rays of the sun. Other examples include the naked thighs and necks of ostriches and the large, heavily vascularized ears of desert jackrabbits and elephants.

We are the hominid analogue of the *Cataglyphis* ant, except that we had a significant internal as well as external heat load and we not only scavenged dead animals as these ants do—but eventually also incapacitated and hunted down our prey.

Mobility generates body heat, and that requires sweat to continue the chase, but you can be profligate with sweat only if you have lots of water. Out of view of the eland or kudu we killed, there was undoubtedly a lake, a stream, or some other water.

Several years ago an issue of *Geo* magazine contained a photograph that is indelibly imprinted on my mind. It shows the bloody mass of a dead elephant whose trunk has been hacked off. Around it swarm more than a dozen armed men who are cutting into the animal with knives and spear blades. The scene is in open bush country, under bright sunshine. African elephants are as naked as men—unlike the woolly mammoths, mastodons, and rhinoceroses, former residents of the northern ice age steppes. Along with those images I also see a petroglyph in a small rock shelter in East Africa showing running hunters chasing a wildebeest, just as the present-day Bushmen chase kudu and run them into submission.

We are not exempt from the physical and biological laws of necessity and constraint that govern all organisms. But this generalization applies especially to our exterior. The remarkable similarity of the DNA signatures of human groups living today suggests that our external differences are trivial; we are all derived from a small founder population, which lived as recently as about 89,000 years ago. We were arguably modest agents promoting species diversity by being responsible for splitting off a second species of louse, where two had sufficed previously and one fewer might suffice even now. However, we were never much "into" tolerance or promotion of diversity. Indeed, there is a puzzling if not disturbing record of disappearance of other human as well as other animal species whenever *Homo sapiens* arrived on the scene.

IN THE NORTH, THE HUMANS COMING OUT OF AFRICA encroached on the domain of the Neanderthal people, who had lived there for perhaps 250,000 years, over three ice ages. And by 30,000 years before the present we had replaced *Homo neanderthalensis*. This northern species was physiologically and behaviorally superbly adapted to a cold climate. Neanderthals had a brain as large as, or possibly slightly larger than, ours. We now consider them not to have been as innovative as *Homo sapiens*, by our standards (which include imagining supernatural beings, drawing pictures, etc.), but there is evidence that they did use fire, decorate their dead with flowers, and perhaps blow on bone flutes. They probably sang, talked, and played. As I will shortly suggest, they were probably also as hairy as bears. They didn't succumb to the weather. They succumbed to something else.

The reason for the Neanderthals' demise is extremely murky, and perhaps that is just as well for our collective ego. But there has been no lack of speculation about what they looked like and how they lived. Primarily, it is thought that since they did not change their stone implements, they were less imaginative than the invaders and would therefore have been outcompeted or killed off or both. I will here add my own two cents' worth, which is not contrary to what has been found and said before, but it adds a zoological twist to the more common paleontological and anthropological perspectives.

First, I will return once more to body hair. If there is one thing which almost everyone agrees on (but for which there is not a stitch of direct evidence), I think it is that Neanderthals were furry. If even some *Homo sapiens* coming north possibly started to become furry (my speculation, from a limited sample of specimens) despite having invented clothes, then the Neanderthals living in the north for 200,000 years or more would have been furry. It might even be a fairly solid inference that they were more furry than any of us are now. Could they have been as furry as other northern mammals? Could they have been as furry as the macaques that have adapted to the cold climate of northern Japan? Fur like that of the northern-adapted macaque or a bear should have made a huge difference to their survival throughout three ice ages, in both males and females. Fur could also have had other implications besides insulation: sexual selection, hybridization or lack thereof, tactical aspects of conflict, and species extinction.

One of the major selective pressures operating on the behavior, physiology, and appearance of vertebrate animals is sexual selection. Sexually selected traits vary enormously, yet they almost always signal something about the bearer that is correlated with survival value and the ability to produce or rear offspring. Survival and fertility markers, or features that can become correlated with strength and vigor—antlers, long tails, and so on—almost by definition become marks of "beauty." The relevant features vary enormously from one species to the next, and what may be very attractive to one should and would probably be repugnant to another. For example, we don't perceive the pale blue scrotum in the background of a bright red penis in the vervet and patas monkeys as a turn-on. The swollen red buttocks of a female chimp seem rather ugly to us, but to male chimps they are a sexual turn-on. If a solid, sleek coat of body hair in Neanderthals had, versus a thin, scraggly coat, survival value, then it would probably become a mark of beauty to them and it would have become more entrenched in their genome as a selected trait. It would be like the *lack* of body fur in warm-adapted *Homo sapiens* hunters. It would be like

Fig. 35. Japanese macaques from the northern part of the island are densely furred, unlike any other living primates.

clothes to us, since clothes have become a necessity for our survival. As a matter of record, we do find body hair a turn-on, or else we would not fuss with it so much. I am talking of course, of head hair, which was probably of great survival value long ago at "crunch" times such as chasing down an antelope in the noonday heat. It may be selected for still, though not for the same reasons.

Large differences in sexual selection signals are especially important in closely related species, where interbreeding is a possibility. The songs and showy feather displays of birds are markers of fitness, but the males of the song sparrow sing entirely different notes and an entirely different repertoire from the white-throated; and the males of one finch species are bright yellow whereas other species are purple or indigo or green. We assume too much when we say that the Neanderthals valued the same things as we did and thus looked like us. External appearance relating to body fur or facial features, or clothes or lack of them, might have accounted for DNA evidence that now indicates there was no interbreeding between us and the Neanderthals. There is reason to believe there might have been outright aggression between the two species.

Neanderthals were probably low-tech survivors, or else more artifacts than crude campfires and scrapers would have been found. Would low technology then perhaps have been compensated for by some other attributes: fur and hibernation? If the Neanderthals hibernated in caves as the northern bears did (and if not, why not?), then we, the summer-adapted human hunters who invaded their territory, would probably have killed a Neanderthal as readily as we would kill a bear. But even if there was resemblance between us and the Neanderthals that could have inhibited our seeing them as prey, that would not have been an absolute deterrent. We are superbly adapted for making "us versus them" distinctions, on the basis of incredibly slight real differences and even on the basis of imagined or created stereotypes. This highlighting of differences is a psychological mechanism that divides, but the capacity to apply it to others probably evolved because it functions within the group to strengthen cohesion for "better" (more efficient) competition against other groups.

I believe the Neanderthals would have been more distinct from us than they seem now from the structure of their bones. They were not inferior. Their apparently more simple lifestyle was probably, given the 200 millennia or more in which it was tested, sustainable and in tune with their environment. In perhaps another century or less we may find out if we can do as well. We will learn if it is possible to live happily and be healthy in mind, body, and spirit surrounded by a devastated fauna, as a perpetual summer species in the north, sustained there by food and energy imported from distant continents.

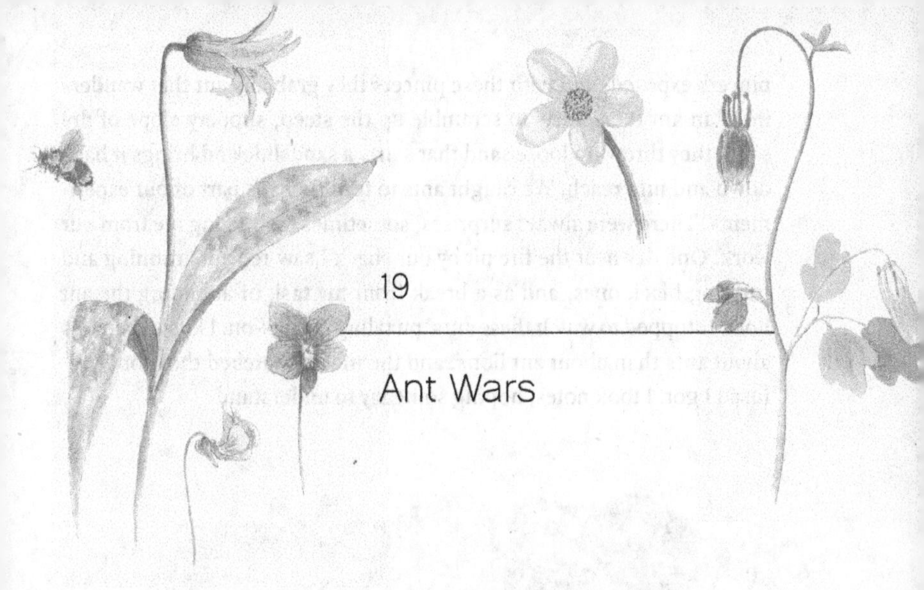

# 19

# Ant Wars

WHY DO PEOPLE WAX ELOQUENT ABOUT THE simple life, especially one spent during the summer in a hut at a lakeshore? Yearning for the simple life in the winter might be less inviting, although if there were a way to induce hibernation—perhaps by injecting the chemical that is produced naturally in bears and woodchucks that hibernate—then it might at least be an attractive option for some obligate summer-lovers. But even in summer, the main problem, if it can be called such, is entertainment or lack thereof, although there is a solution that beats a lawn mower, a lawn chair, or a television set with 100 channels, by a mile: watching ants and other critters.

EVERY SUMMER I SPEND SOME TIME TRYING TO LEARN something new about animals. I spent the summer of 1981 in the Maine woods, living in and out of an old, dilapidated one-room tar-paper shack with my wife then, a great horned owl, and two crows. Maggie and I were studying the behavior of insects commonly known as ant lions because they are slow-moving predators that catch fast ants. They do this by making pits in loose, dry sand. The pits serve as traps; the ant lions hide buried in sand at the bottom of the traps with only their sharp tonglike pincers exposed; and with these pincers they grab any ant that wanders in. If an ant then starts to scramble up the steep, slippery slope of dry sand, they throw up loose sand that starts a sandslide and brings it back down and into reach. We caught ants to feed them as part of our experiments. There were always surprises, sometimes distracting me from our work. One day near the fire pit by our shack I saw red ants running and carrying black ones, and as a break from my task of attending the ant lions I stopped to watch these ants' puzzling goings-on. I knew even less about ants than about ant lions, and the more I watched the more confused I got. I took notes, hoping someday to understand.

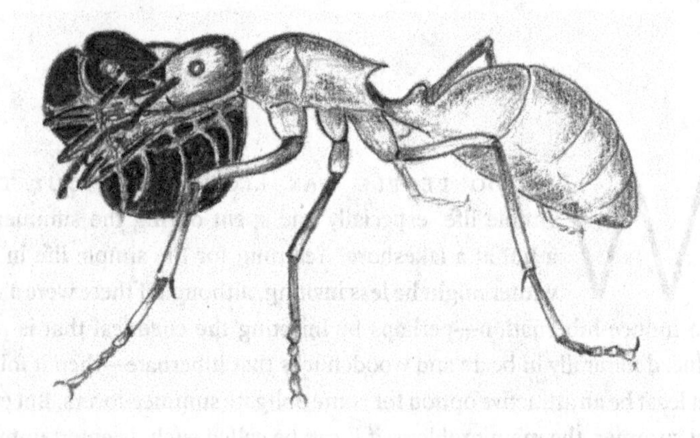

Fig. 36. An ant carrying an apparently very willing one from another species, who remains totally motionless.

Next to our fire pit was a black ant mound that was then a shallow pile of loose soil, balsam fir needles, and other debris. Hairy-cap moss grew on and around the old ant heap, and wild blueberries grew next to it. The colony was, I thought, being invaded by a column of big red ants that were taking out the black ants and lugging them off over the glacier-grooved granite ledge of our doorstep and through a low-bush blueberry patch. I traced them to a nest mound at the edge of the pine forest nearly

100 feet to the north. I assumed that it was a slave raid. Ant "slaves" result from immatures (generally larvae or pupae) that are taken from another nest, and that then acquire the odor of the colony into which they are brought. After aquiring the colony odor, they are accepted as colony members. But I was puzzled to see no fuss and fighting among the adult ants; each "slave" curled itself up into a little ball to be easily carried. I wondered why they went so willingly.

I WAS TO COME BACK FOR A FEW MORE SUMMERS. DURING the next summer, 1982, the same colony of red ants "raided" two more black ant nests, one of which was in another clearing at an impressive distance: 250 feet. To reach this nest the reds had to traverse a shady spruce-fir thicket. I speculated that what I was seeing was a regular occurrence, because there were twenty-one empty ant mounds well within range of the big mound of the red raiders. I knew from their defensive smell when I disturbed their nest that these were ants of the genus *Formica*, or formic acid ants. In Europe, red *Formica* wood ants control caterpillar outbreaks. A single colony can reputedly dispatch 100,000 caterpillars per day. Perhaps these ants here, which looked very similar to me, were concentrating on other prey, since caterpillars were only rarely dragged into the nest.

On 11 August a column of red ants traveled to the same distant north colony as the previous summer, and I took some notes while watching their trail. During a ten-minute period I saw ninety-one red ants pass by carrying ant brood (six large larvae or ant grubs and eighty-five pupae), and forty-seven were carrying adult black ants. As before, there was no hint of a struggle; each of the captive black ants seemed to tuck itself into a little ball, the better to be carried.

The carriers were not significantly slowed. Unloaded ants were passing at an average pace of 1.9 inches per second, whereas those carrying another ant were running at 1.7 inches per second. The apparent captives held still for the entire time—approximately half an hour—required for them to be carried the total distance of 250 feet from their nest. Before, I had seldom seen an ant stand still for even a few seconds. Were they drugged? To find out, I caught pairs and released several captives from the grip of their red captors, and these instantly ran off as fast as ants run. They were obviously in great shape. Why did they never run off on their own?

The multi-day raid (as I took it to be) always stopped in the evening and resumed late the next morning. This pattern continued for five days, and I started to notice other odd details. Occasionally, a black ant carried another black one, and then—even more puzzling to me—I saw a black ant carrying a red one. (The two differently colored ants were *Formica subintegra*, the red; and *F. fusca*, the black). I didn't know what the anomalous behavior meant, but I brushed it off as typical ant confusion, rather than ignorance on my part.

Strangely, I saw no battles at the raided nest. I did see one red ant tussle with a black ant, but the latter turned belly up surprisingly quickly and then submitted to the tuck position so it could be carried away. Would it have been killed if it had resisted?

Hoping to get to the bottom of the ants' strange behavior, I finally dug into the red ants' nest where the blacks were being deposited. To my now ever-increasing surprise it didn't look much like a red ant mound at all. In fact, inside this nest the blacks outnumbered the reds. In a random count, I tallied 178 blacks and only 23 reds. Was this actually a black colony rather than a red colony? Five days later I was again watching the continuing drama as large numbers of reds left their fortress to hit the trail and head north. I counted fifty-six reds carrying blacks, seven reds carrying reds, and one black carrying a red. So, proportionally at least, the reds were the main carriers. Along the same ant column I saw two black queens being pinned down by six to ten red ants each. The next day (17 August 1982) the raided mound was nearly empty. But in the morning the knot of reds were still pinning down a black queen (the same one?) on the trail at the same spot where they had been holding one yesterday. And by early afternoon that queen was still being held by a mob of about fifty reds. Why didn't they kill her?

I surveyed the ant mounds in the clearing around our shack, and then also at a larger clearing nearby where we were building a log cabin. I found that the reds were in the minority: thirty-nine of forty-one colonies were populated exclusively by blacks. The reds were not just raiders; they had other professions as well: fifteen of seventeen aphid colonies on young poplar saplings were tended by them, and I found only two of the blacks. Additionally, both kinds of ants tended little green nubs that looked like aphids, on the petioles of new chokecherry leaves. The ants must have been getting some secretion from these growths and using it as a guard to keep off caterpillars that might otherwise eat them: I put a caterpillar on a chokecherry twig with ants—it was instantly attacked; the green nubs on the leaf petioles were probably an adaptive design.

For four days in late August a steady stream of black ants carried in other black ants from a colony about forty feet distant. I had no idea what this meant, except that another summer of watching ants was obviously needed, and eagerly awaited.

In my third summer, 1983, I again kept a sharp eye out for the ants. To my great surprise, on 14 May when I first came and looked, I saw only one red ant for about every 100 black ones. Just to be sure, I counted again three days later and got exactly the same result: 396 blacks, four reds. The black ants were busy carrying soil out of the mound and fir needles onto it. The very few reds were not idle; they were carrying debris onto the mound as well, and I saw them help blacks drag in a dead fly and a caterpillar. Curiously, on this day I also saw one big black ant carry a smaller black ant over the top of the mound, and yet again another black one carrying a red tucked up in the typical carrying posture. There was no raid in progress. Who gets to be carried, and why?

Perhaps I could at least use an experiment to find out who would raid whom. I needed to confront one ant colony with another to find out who did what. In July I dug up a colony of black ants; put it into a bucket; and dumped it, brood and all, about six feet from a nest populated by blacks and also reds. The result was almost instantaneous, and dramatic. Within minutes the presumably disorganized, weakened black colony was attacked by thousands of reds swarming onto and into it from their settled nearby nest. They took the black ants' brood and carried it back, but these red raiders brought *no* black adults back.

These observations suggested to me that the blacks I had previously seen being carried by reds were individuals who had emerged in the reds' nests from brood that had been carried in previously. The colony odor is like an identity badge. The blacks that emerged from these pupae had taken on the colony odor, had blended in with the reds, and had become indistinguishable from them—ants may be color-blind, but they are not scent-blind. Therefore, the "slaves" that were carried in are, after they emerge from pupae, at least theoretically perfectly situated to exploit the reds, their hosts. I could hardly wait to find out if they might produce sexual adults (alates, those who begin life having wings for dispersal to start their own nests) in their hosts' nest, as might be possible and even likely unless the raiders selectively kill queen pupae.

Having raided the weakened black colony, the reds had apparently whetted their enthusiasm for more. On the very next day, 14 July, they were already on the march again, this time to another black nest, an intact though small one. As before, this time there was not a single black ant being carried back; only their pupae were being carried. I wondered if these could have included pupae not just of workers but also of potential drones and females (queens). If the carriers didn't make this fine distinction between sterile workers and reproductives (who do no work in the colony where they are born), then taking "slaves" could have a cost, since any reproductives of the other species would simply leave the colony and provide no labor.

The blacks being raided by the reds were also carrying brood, but they were running off in the opposite direction, carrying their remaining brood. There was fighting at the mound, and dead bodies and body parts were liberally strewn around. This, then, was a definitive slave raid. (I saw the red ants make others in identical manner on subsequent occasions.)

By 25 July a nest of reds, which I had seen making at least two "slave" raids on blacks that summer, was again showing lots of traffic to what appeared to be the main portion of its colony living in a separate mound at the edge of the clearing. Now I saw both blacks and reds carrying brood and adults, as before. That is, I realized this colony of both red and blacks was dispersed into at least two domiciles, between which it shifted its colony members (much as we move from home to camp and back again, depending on the season or the weather). Four days later, on 29 July, while the colony transfer from one nest to the other was still in progress, I dug up the satellite nest. Here, finally, I saw what I was looking for: *winged* ants (i.e., the virgin reproductives). I counted 154 males and ninety-five females (queens). Winged males and females eventually leave their parent colony and disperse in all directions. Back on the ground after being mated with the males from other colonies, the females then break their wings off and settle down to commence a lifetime of egg laying.

31 July 1983. It is a beautiful sunny morning with no wind, and the screen tent that I previously placed over the main colony of the reds has finally paid off. Alates flew up into the tent, where I could retrieve them (as I had hoped, since I expected them to go to the light in the sky for their nuptials and their dispersal). That morning I captured a flight of 194 males. (It made sense for the females not to swarm simultaneously, to reduce inbreeding.) None left during the rest of the day. On 3 August the nest issued at least twenty-five more males, and on 8 August another 100 males came from the main nest.

Now came the rub. I knew that the workers are smaller-bodied versions of queens (except that virgin queens have wings), so the new queens should be easy to identify. But I had no idea what males might look like. These males were all black. Could they potentially be males of the black species *Formica fusca*?

Ant taxonomy is a difficult subject, and I was stuck. This was not something I could solve by observations and experiments, so I went straight to the authority: I sent them to the premier ant specialist, Edward O. Wilson. Contrary to what I might have naively assumed from superficial appearance, the black male ants were identified as "reds," *F. subintegra*, just as they were supposed to be according to standard ant lore. (I later had occasion to dig up a nest of the black ants, *F. fusca*, and found some of their males ready to leave. They were also black, but these had red legs and darkly pigmented wings.)

Bert Hölldobler, another world authority on ants, wrote to me:

> You observed raids, but you also observed nest emigrations. During raids, *only* pupae or fully grown larvae are taken by the raiders. When emigrations to another nest occur, then the younger workers are also carried. You observed an emigration of the mixed (slaves and raiders) colony where mostly the raider species was seen as carrier. This was probably due to the fact that they were the older workers, and the black ones (the slaves) the younger. Emigrations occur especially in late summer and fall,

when many ants shift nest sites, because many species propagate by budding or establish "winter nests."

I had not made original discoveries, but no discoveries can be made without exploring, and thanks to my ignorance I had been lured to try. I had fun, I had learned much about ants, and they had helped make several summers special.

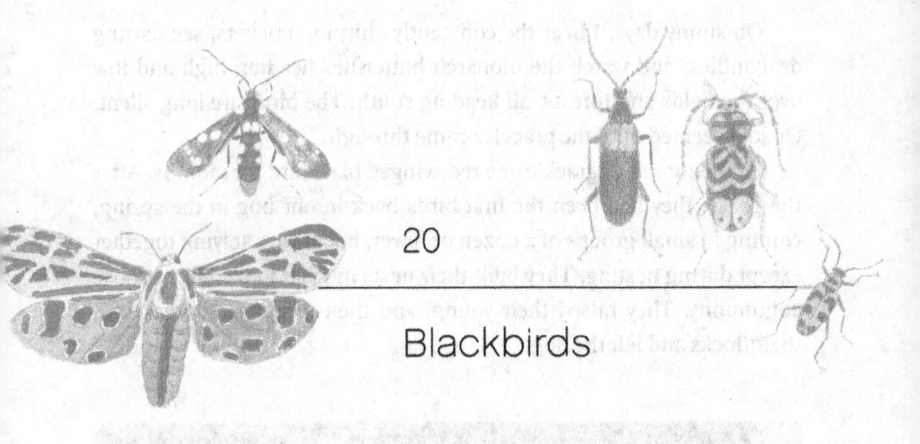

# 20

# Blackbirds

24 September 2005. MOST TREES ARE STILL BRIGHT green, but the forest is now becoming a rich palette where individual trees have definition, as those still green contrast with others—the black-greens of the balsam firs, and here and there some gold and orange and a few points of bright red brilliance from the maples. The colors are most impressive when the sky is leaden, when drifting clouds bring diffuse light that illuminates the colors—bright sun bleaches them out.

Caterpillars of many kinds are common this fall. Now, in late September, is a good time to find some of the big moth caterpillars. Since many of the birds are already gone, perhaps they are now safer from birds—but they are never safe from parasites. I have kept track of a waved sphinx moth (*Ceratomia undulosa*) caterpillar on an ash sapling as it fed on leaves, trimming them down along the edge instead of leaving conspicuous holes, and then clipping off the remaining uneaten portions. I predicted it would *not* clip off the last few leaves before it left the plant to pupate. Suddenly it stopped feeding entirely, and for a couple of days it just hung from a leaf in the "sphinx" position. One morning its green skin was covered with ninety-one freshly spun white braconid wasp cocoons. Its skin had little dark puncture wounds where the larvae had poked (chewed?) through to come out.

On sunny days, I hear the constantly chirping crickets, see darting dragonflies, and watch the monarch butterflies heading high and low over the fields and forests, all heading south. The birds are long silent. Or so it seemed until the grackles came through.

I have not seen a grackle or a red-winged blackbird for months. After the geese, they had been the first birds back in our bog in the spring, coming in small groups of a dozen or fewer, but always staying together except during nesting. They built their nests in what looked like a loose community. They raised their young, and then disappeared back into their flocks and left the bog.

Fig. 37. **Braconid wasp cocoons spun by larvae freshly emerged from the sphinx moth caterpillar on which they are loosely attached.**

I wondered where they went. Today, on my jog through a wooded pass near a cliff where ravens nest, I found out. I came on a swarm of thousands of grackles. Long before I got close to them, or vice versa, I heard their ruckus of squeaks, squeals, screeches, and nasal "ticks," "tucks," and "tocks" that together made a roar. A broad black stream of the birds braided through the branches of the maples, oaks, birches, and cherries. I stopped and stood still, mesmerized as a river of them passed over and around me. A crowd would build up in one tree; others would come to join it; more and more would join the crowd; and suddenly they all lifted off in a roar of wings accompanied by momentarily silenced voices. The roar would subside and the chatter resume at another tree. Black streams of them kept flowing by, sounding like wind blowing through trees. In a few moments any one stream would be reduced to a trickle, and then the volume of the flow would pick up again.

The birds of this grackle aggregation were all foraging in the treetops, where many of them were tearing open silked-together maple leaves with little green caterpillars inside. I had seen a similar swarm of grackles several years earlier, but later in the season—perhaps October or early November, after the leaves had fallen. Those birds had foraged on the ground in the forest and they proceeded like a giant wheel, with those on the advancing front flying ahead of those already on the ground, then falling behind as the ones in the rear overtook them, to again fly ahead. And thus they advanced through the woods like a vacuum cleaner, presumably sweeping up food as they went.

Grackles are highly social animals, and as such they have a "personality" that we find appealing; they cozy up to each other, or to us if we become part of their group. Personality shows up through contrasts, and nowhere was the difference more obvious than between a young robin and a young grackle that we raised at the same time. Our pet grackle, named Crackle, followed us—especially when he (or she) was hungry; he would then come instantly if we called him by name. Crackle even followed us into the house; and he banged against the window a couple of times trying to get in, one time hitting it so hard that he got knocked out—we were thankful he revived. He also begged at the door whenever we made some commotion inside. The robin, by contrast, perched on the porch under the branches of our shading cherry tree, and there it stayed and peeped. It seemed to be rooted in place as though wanting us to come to it. The robin gaped in our presence, but it would orient itself only to the food. Crackle looked at us and oriented to the person.

The robin quickly searched for and found worms. Crackle found very little on his own, but he picked up and examined all odd objects we showed him. The robin was hardwired, fixated on worms and uninterested in anything novel. Both birds ate worms, but the robin swallowed them expertly. As soon as it got even just the tip of a worm in its bill—bam, it got the whole worm down the gullet. Crackle had to wrestle with each worm, and as often as not the worm escaped from his bill. Crackle was quick to hide under a tarp in the rain. The robin hunkered down in the rain and got soaked.

Grackles must have some good reason to travel (and sleep) in huge crowds. Aside from the trivial proximal reason that they like company, what is in it for them? What is the selective advantage that makes the grackles want to be associated with others? There are many possible mutually nonexclusive reasons, such as safety in numbers; sharing information so as to find food, identify an enemy, or issue a warning; and better access to food (such as by flushing prey). But I doubt that these late-summer foraging crowds flush much prey for each other; the insects they forage from don't fly or drop from the leaves. Better access to food, such as raven crowds gain by overpowering strong defenders, is also not a real option. Mutual education—showing each other where to find food—seems possible, but more likely they also compete for food. Only one thing is certain—they already live in a different world from the one they inhabited all summer, even though they have not yet even left the state.

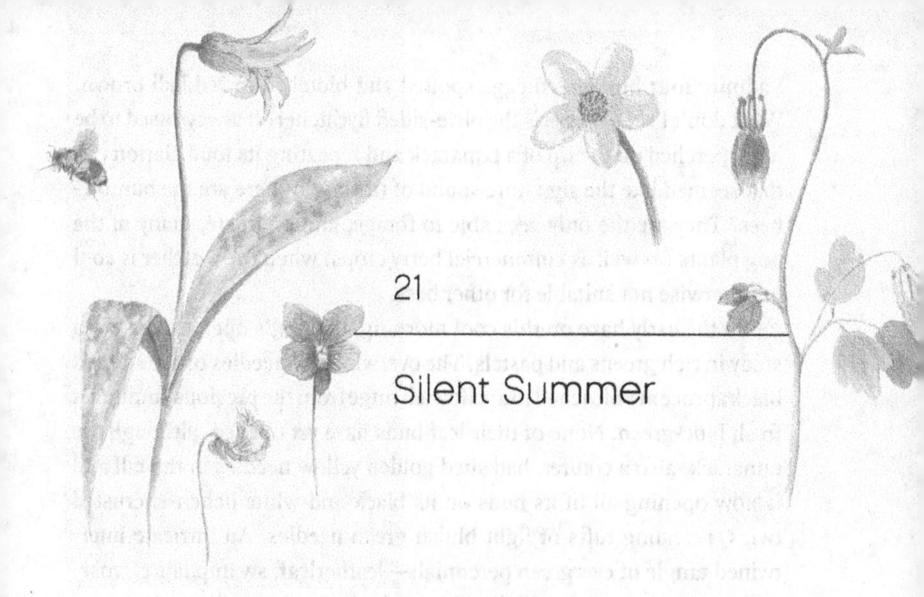

# 21

# Silent Summer

I BREW MYSELF SOME COFFEE AND THEN LEAVE TO revisit Huckleberry Bog, a haven for many plants and animals that are not found in the forest. It is surrounded by forest and edged by a cordon of dense brushy thickets growing in algae-covered water. I don tough pants and a shirt when I need to force my way through, but I don't wear boots—they get filled with ice-cold muddy water when I get to hip-deep holes or beaver channels. Wet cold feet are obligatory, and I put on worn-out running shoes. With some relief I finally break through to enter the bog proper, where I am in the open and walk on an ancient mat of roots and sphagnum (peat) moss that has grown over a glacial pond. Some of the same tree species that grow at the periphery are also present—red maples, black spruce, and larch. But here they are scattered and look stunted, like bonsai. As if I were walking on a waterbed, each footstep depresses the root and moss mat, and it sinks a few inches, to rebound when I lift my foot—hence the name "floating bog." Somewhere still preserved in the solid bottom below me is the pollen of plants that grew on surrounding hills after the last ice age. Did a woolly mammoth or two break through and leave its bones here also? Except for the refrain of the yellowthroat and six other birdsongs, the bog remains silent. It does not tell. But a white-throated sparrow flushes near my feet, and I gaze into its nest cup, which is sunk into the wet moss.

I admire four blue-green eggs spotted and blotched in reddish brown. What don't I see? Where is the olive-sided flycatcher? It always used to be here, perched on the tip of a tamarack and repeating its loud clarion call that seemed like the signature sound of the bog. Where are the bumblebees? They are the only bees able to forage, and pollinate, many of the bog plants (as well as commercial berry crops) when the weather is cool or otherwise not suitable for other bees.

In the early haze on this cool morning, the bog's open expanse is a study in rich greens and pastels. The overwintered needles of the stunted black spruce are bleached to a yellowish tinge from the previous summer's fresh blue-green. None of their leaf buds have yet opened, although the tamarack, also a conifer, had shed golden yellow needles in the fall and is now opening all of its buds on its black-and-white lichen-encrusted twigs, revealing tufts of light bluish green needles. An intricate intertwined tangle of evergreen perennials—leatherleaf, swamp laurel, rosemary, Labrador tea, lambkill, and cranberry—rests on the sphagnum, and as my feet sink in I see small clumps of bright pink blossoms of the swamp laurel, and shining white ones of the rosemary. At the water along the edge grow the taller and deciduous plants—high-bush blueberries, huckleberry, winterberry, chokeberry, and privet andromeda—all now putting on new yellowish to bluish green leaves. The color combinations of the leaves, buds, twigs, flowers, and berries—greens, browns, yellows, red, gray, black—are artistically perfect. I pick a selection of the plant riches to take home so that I can make a token sketch. It can be only a reminder of the beauty and perfection of this place, a piece of work of the Creation.

Bumblebee queens have been out of hibernation for at least two or three weeks, and by now they would have found nest sites and would be starting their annual colonies. All the flowers that are concentrated here should be a magnet for them. But today, like the last time I was here, I hear and see almost none. Even after slogging all around the bog until my feet are numb from the cold, I again see none of the many species expected. I see only one *Bombus vagans* and one *B. ternarius*, the latter a pretty black, yellow, and orange bee. Has something happened to the bees?

One species, *Bombus terricola*, which used to be the most common one from my clearing in the woods to the tops of the nearby mountains, into this bog, and into the woods in the northern Maine wilderness, seems to be totally gone. I have not seen it for years, and I am again shocked not to see it today. I have seen few bees of any kind, however, and I am not yet worried, because bumblebee populations, like populations of other social insects, such as wasps and hornets, keep growing throughout the season. Each queen will produce hundreds of workers as the summer progresses. By July the population will appear to have dramatically increased, since the queens who before then spent most of their time hidden inside nests where they incubated their eggs and larvae have produced crowds of workers and drones. Thus, late summer is the best time to see which species are locally present. *Bombus terricola* was still present, though very rare. During two years' search I ended up seeing three workers in Maine and one in Vermont.

On this day in May the bog looked pristine and nothing seemed changed except for the apparently total absence of a species that hardly anyone would have looked for, or noticed. What happened?

SOME YEARS AGO I FOUND ABOUT A DOZEN MOUNDS OF Styrofoam chips mixed with peat buried in the bog to grow something above the water level. I think they were the remains of potting soil for marijuana plants that someone had grown in the most hidden spot he or she could find. The foreign plants had long since been removed, but I was appalled to see a substance in this natural ecosystem that did not belong here. I spent half a day digging it up; hauling it out through the woods and up to the road; trucking it away; and paying to leave it at a dump, even though I was trespassing myself in the bog (since I had no idea who owned it). This time the bog was apparently no longer being used as a dump for piles of Styrofoam (mixed with soil), but on my walk I had passed a real dump site on the side of the hill toward the bog, and I was again shocked, angry, and—more than anything else—also afraid. This dump site contained an unsightly collection of plastic, other discarded petroleum products, dozens of tires, and other detritus. Could poisons be released from any of these products of our chemically synthesized civilization to bio-accumulate and disrupt the metabolism of an ecosystem?

Any foreign chemical put into the ecosystem, whether the woods, a swamp, or the body, is by definition guilty until proved innocent. And innocence is difficult to prove, since effects may be slow, may be long delayed, and may pop up in the least expected places. I am talking about natural versus unnatural compounds, although I do not mean that natural compounds are nontoxic. On the contrary, some of the most toxic chemicals known to man are produced naturally by plants and animals and usually serve as a defense. These need to and do have immediate and all too painfully obvious effects, but they do not accumulate in the environment as plastic does.

There is an incredible toughness to life, but at the same time a frailty that borders on the absurd. We supplement our milk with vitamin D, which we need for our health, but vitamin D is also used as a rodenticide. A polar bear has a huge amount of vitamin A in its liver, enough to kill a human who eats it. I think I have been especially sensitive to frailty versus toughness because of my experience in trying to keep wild animals as pets or to domesticate wild plants—to say nothing about trying to get them to reproduce. Almost invariably, for any one species there is a huge list of what seem like absurdly fussy requirements related to its natural environment, requirements that are often almost impossible to consciously duplicate.

One chemical, which seemed so nontoxic that people offered to ingest it, turned out to be deadly. This was dichlorodiphenyltrichloroethane, more commonly referred to as DDT, an insect poison that broke down to dichlorodiphenyldichloroethylene (DDE), the active culprit that affected birds. It took years for the effects of this toxin to be identified. The offending chemical first had to be passed on from one organism to another; in some of these organisms it did no discernible harm, but eventually it reached some which it did harm. It was slow-acting and did not directly affect the birds' mortality. It affected their behavior, and the thickness of their eggshells—especially in pelicans, who ate fish that fed on aquatic invertebrates; and in raptorial birds, particularly falcons, because they ate birds who had eaten insects. Thanks to the alerts sounded by naturalists, and to long, patient, expensive sleuthing, DDT was eventually identified as the source of the devastation. Heroic countermeasures were instigated, and they reversed the trend toward what would otherwise have been the obliteration of more than I dare to contemplate. The scary part is that DDT was solemnly sworn to be a safe chemical—it had been extensively tested before being released. And now, many decades later, we are still finding out more: for example, that exposure of girls to DDT prior to puberty greatly increases their risk of breast cancer later in life. We still release about fifty new chemicals into circulation per week. They are tested on lab rats—animals that never experience summer or winter, that live well in dumps, and that when tested have no relation to any ecosystem except a sterile cubic plastic box. The chemicals don't get tested in a pristine bog where the olive-sided flycatcher sings and the bumblebees collect nectar and pollen from pink rhodora blossoms in early summer, and where blue pickerelweed flowers poke up out of the water of the languid stream flowing through it in July and August.

In the summer of 2008, I finally saw *Bombus terricola* again. I found one dead in Hinesburg, Vermont, and in Maine I regularly saw several live ones in three places where I looked (Hog Island, in Muscongus Bay; in western Maine on my hill, and near Orland). A continuing comeback of the species is likely. I suspect now that its severe setback for over two decades could have been due to a "wildfire" effect; a very high former population was dense enough for an emergent or new pathogen to easily spread from one bee to another. The high bee population favored the lethality of the pathogen on those bees. If this is correct, the surviving bees will now have evolved increased resistance, and the surviving pathogens will have evolved reduced virulence. A dieback by chemicals, on the other hand, would likely have affected many species simultaneously.

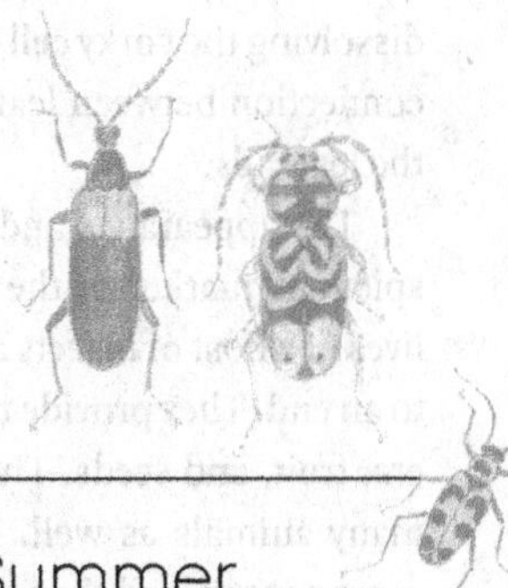

# 22

# Ending Summer

CHEMICALS CAN BE POWERFUL. ONE—THE CHLOrophyll molecule—causes the greening in spring that is the basis of the summer world. Without chlorophyll, life as we know it would not exist on Earth. Another chemical, which trees also produce, arguably causes the end of summer, because after it has done its job the caterpillars are gone and most of the birds must leave. This molecule, abscisic acid, releases the leaves' hold on twigs and causes defoliation at summer's end.

Just as important, of course, are the cues that cause plants to produce these key chemicals. Warming in the spring brings trees to life; but lower temperatures alone do not cause a tree to shut down and lose its leaves at the end of summer. The relatively precise timing of the leaf fall is of some importance, and it has been "calculated" by evolution through a balance of costs and benefits. Photoperiod, specifically the length of nights, is the main stimulus.

When nights become long enough, trees begin to shut down for the summer by forming a corky layer of cells between leaf and twig. This layer, the abscission layer, then blocks off the transport of materials between the branch and the tree. Chlorophyll is then no longer replaced as it breaks down with use; and as it disintegrates, the yellow and orange leaf pigments are revealed. The abscisic acid then does its job of dissolving the corky cell layer that holds the leaves to the trees, and as the connection between leaf and twig weakens, a breeze does the rest, and the leaf falls.

The appearance and disappearance of leaves may be the most conspicuous marker of the seasons and the most important events to the lives of a host of insects and birds; but to the plant, leaves are only means to an end. They provide the energy and raw materials for producing flowers, fruit, and seeds. The timing of the bloom and fruiting is critical to many animals as well. The summer schedule for the tree's leafing is more constrained than that of the bloom, which may appear to be totally random, since it starts long before the leaves come out in some species, occurs in midsummer in others, and extends until late fall in at least one species, witch hazel. But it's not random at all. All the deciduous trees (those that shed their leaves in the fall) that are wind-pollinated flower before the leaves come on, probably because leaves would hinder access of airborne pollen to the pistils. Wind pollination, as such, does not require early blossoming, because all the wind-pollinated conifers flower not in early spring but in the summer. Bee-pollinated flowers, such as those of black locust and basswood, flower late in the summer, when insect pollinator populations have built up. Witch hazel, which flowers in September, is pollinated by winter moths that are beginning to be active only then. Animal-pollinated trees bloom when pollinators are available, so in theory they could be pollinated all summer long, except that even in an undisturbed habitat there is competition for pollinators. Fewer pollinators are available to any one tree species if another one that blooms simultaneously draws some away to its flowers. However, divergence of species' blooming times—so that these times space themselves out over the entire summer—is one of several solutions that reduce competition.

The timing of blooming is also tactical in part because it secondarily affects the timing of fruiting. Different species of bamboo, for example, flower and produce large seed crops not once per year, but at intervals of 60 to 100 years or more. Furthermore, when they do flower, they do so synchronously over vast areas. The naturalist George B. Schaller noted that in 1974 and 1976 umbrella bamboo—a staple of pandas—died throughout an area of 2,000 square miles in the pandas' northern range. At least 140 of the rare pandas died. Undoubtedly, vast numbers of ro-

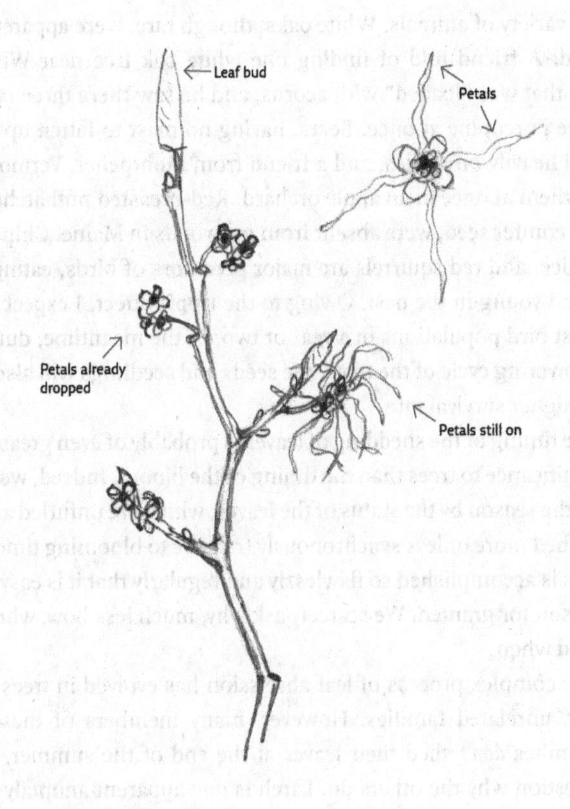


Fig. 38. Witch hazel flowering in October.

dents—seed predators of the bamboo—died as well. If the bamboo were to flower and produce seed every year, the rodent populations would be permanently high and perhaps harvest all the seed produced each year. Similarly, some tree species in the woods from Maine to Vermont also time their blooming by *not* blooming, and thereby control the seed predator populations. In the summer of 2007, for example, the sugar maples, American ash, red oak, beech, white pine, and red spruce all failed to flower, and as a consequence there were almost no sugar maple seeds, acorns, or beechnuts—a collective absence of the mast that feeds a large variety of animals. White oaks, though rare, were apparently not affected. A friend told of finding one white oak tree near Wiscasset, Maine, that was "loaded" with acorns, and he saw there three raccoons and one porcupine at once. Bears, having no mast to fatten up on, depended heavily on apples, and a friend from Montpelier, Vermont, saw five of them at once in an apple orchard. Red-breasted nuthatches, who rely on conifer seed, were absent from my woods in Maine. Chipmunks, deer mice, and red squirrels are major predators of birds, eating birds' eggs and young in the nest. Owing to the ripple effect, I expect a surge of forest bird populations in a year or two. In the meantime, during the next flowering cycle of the trees, the seeds and seedlings will also have a much higher survival rate.

The timing of the shedding of leaves is probably of even greater tactical significance to trees than the timing of the bloom. Indeed, we almost define the season by the status of the leaves, which are unfurled and then again shed more or less synchronously (relative to blooming times). The process is accomplished so flawlessly and regularly that it is easy to take the reason for granted. We scarcely ask why, much less how, whether or not, and when.

The complex process of leaf abscission has evolved in trees of very diverse, unrelated families. However, many members of these same tree families *don't* shed their leaves at the end of the summer, raising the question why the others do. Larch is one apparent anomaly among the conifers. Its leaves turn golden in the fall and are all shed before winter, having served the tree about five months. The white pine sheds its leaves only after two years, so half its leaves are shed each year. Spruce and balsam fir may keep their leaves for five or six years. Most northern broad-leafed trees shed all their leaves every fall, though some of the more southern trees, like magnolia and some oaks, may keep them not just for an entire year but for five or six years. Since leaf shedding evolved numerous times, it must have offered a powerful selective advantage. But what is that advantage?

The advantage of *not* shedding undamaged leaves, all other things being equal, seems obvious. Leaves are solar panels, and discarding them every few months means having to make new ones later, using time and resources that would otherwise be available for growth. Resources invested in disposable leaves would be valuable for more growth and thus for survival in the fight for light that most forest trees wage against each other throughout their lives. These resources would also be valuable in obtaining a surplus of energy for fruit and seed production. All else being equal, it should be more economical to retain leaves for a whole year, or preferably for several years, than to discard leaves used only for four months and construct new ones every summer. Retaining leaves can serve the added advantage that they are then available for use during the occasional warm spells that almost invariably occur every winter. And as might be predicted from this rationale, many trees (as already mentioned) do indeed keep individual leaves for several years before finally replacing them.

One hypothesis regarding why trees shed leaves before winter is that the leaves would be or are killed by freezing and then are shed incidentally. But this hypothesis does not satisfy the ultimate, evolutionary, question. Frost intolerance of those leaves that are normally shed may be a proximal result of not experiencing freezing and therefore not having had a need to evolve frost-hardiness. By contrast, buds (which contain embryonic stems, leaves, and flowers) are frost-tolerant, even on trees that have frost-sensitive leaves.

Frost-hardiness has evolved in the leaves of many trees. Spruce and fir leaves, for example, withstand temperatures as low as nearly minus 80°F, and they are retained even at the northern limits of tree growth. Broad-leafed trees that grow in north temperate areas where there are commonly winter frosts and that nevertheless keep their leaves alive all winter include some species of each of the following: oaks, hollies, magnolias, rhododendrons, and viburnums. Of course, in the moist lowland tropics, most broad-leafed trees hold their leaves for many years, although in areas subjected to regular drought they shed the leaves, presumably to reduce water loss.

There must be a good reason why many northern trees shed their leaves whereas others keep them. My hypothesis regarding which trees do shed and which don't depends on a conflict of selective pressures that trees must face in the northern hemisphere in areas where there is a lot of precipitation: a large leaf surface area is needed to intercept solar radiation and to absorb carbon dioxide in the summer; but the same leaf surface is a liability in the winter because snow loading could collapse the tree.

What we observe now is a result of evolution over hundreds of millions of years. But the selective pressures that have acted on some features in the past are now unlikely to occur every year and may be seen only rarely. Instead, they are probably witnessed only at bottlenecks. One such event occurred in New England on 26 October 2005, near our home in Vermont. The following journal entry for that date describes what happened:

> It rained all day yesterday, and temperatures were dropping gradually: to 40, 39, and 35°F by evening. The sky stayed dark. Flocks of geese passed over. I woke up in the dark, and the light switch did not respond. I then looked out—SNOW! I went back to bed and waited for daylight before brewing a cup of coffee and stepping outside for a closer look. It was still snowing, and the outside thermometer then read 29°F. I saw devastation—the result of a confluence of rather precise temperature changes, wind directions, clouds, and all this weather in relation to the timing of the leaf shedding of the trees. A perfect timing, complete with proper experimental controls, had produced a rare natural experiment.

At that time, near the end of October, many trees—red oak, quaking aspen, apple, black locust, and silver maple—still retained their full complement of green leaves. Other deciduous trees, including white ash, elm, and red maple, had lost all theirs. Some of the sugar maples, black cherry, and white birches were bare, but a few still had branches that retained most of their leaves, by now golden.

The effect of the snow on individual trees was dramatic but unrelated to the species as such. Trees that retained their leaves paid a steep price. Those that had shed their leaves suffered no damage. The thin, young maples and oaks in the woods around our house were snapped in half or bent to the ground. Similarly, old sugar maples with heavy trunks had huge limbs broken off, and many of their other limbs were bent and ready to snap. The black cherry next to the house had retained its leaves;

and while I was getting wood out of the shed for our stove, three of its huge limbs cracked and fell. One after another, they came crashing to the ground. From the nearby woods, I heard what sounded like muffled rifle shots followed by dull thumps: tall poplars were falling. The red oaks that had suffered the least damage during the great ice storm of 8 January 1998 were now hit the worst. Healthy oak trunks a foot or more in diameter had bent and shattered. Limbs two or more feet thick lay heaped on the ground. In sharp contrast, *no* trees or limbs that had shed their leaves (these trees included some maples, cherries, poplars, and oaks) were damaged. As I later learned, the same scene was enacted over a large part of northern New England, especially at the higher elevations.

Trees face the same problem of timing at the beginning of the summer, as was shown at the same location on 30 May 1996. As is typical by the end of May, all the trees had just fully leafed out. That night, it started to rain, and it got colder at the same time. By nightfall, temperatures dipped slightly below the freezing point, and the rain turned to snow. As more snow fell, it stuck to the already wet leaves and froze on. Throughout the night, temperatures continued to hover around the freezing point, so that the snow did not melt but was wet enough to stick. By morning, critical snow loads had accumulated, and I heard loud crashing throughout our woods as huge limbs snapped and came thundering down. All the large-leafed (deciduous) trees were damaged, but no trees and branches that had no leaves, and none with needles (small, nondeciduous leaves) were affected. Half a foot of snow on the ground at this time certainly did not seem usual, but it may not have been unknown to the 200-year-old trees. It would have been common in the long evolutionary history that shaped them.

These natural "experiments" demonstrate what may seem obvious: trees need solid scaffolding to hold their leaves up to the light. The costs and risks that are involved are a necessity resulting from competition. Most mature forest trees have dropped millions of viable seeds and perhaps produced thousands of seedlings in their lifetime, but of course on average, in a stable population, only one seedling can grow into another tree to replace the parent. The seedlings derived from any one tree are in an intense race to put on growth, and only one of them may grow large enough to break through the canopy and capture enough solar energy to produce seeds as well. Given such competition, one might suppose that trees would continue growing to the very end of summer. Instead, stem growth is usually completed in June, near the midpoint of summer when the ends of the twigs become capped with buds and stop growing. The buds then stay dormant through the warm weather of late summer, through the fall, and through the winter. Generally, they become ready to unfurl their leaves and flowers only in the warmth of the next year. Why do the trees stop growing taller with at least three months of warm weather still to go, long before low temperatures can put a damper on further growth?

I do not have an adequate answer to this question. However, I speculate that backup support has something to do with it. I have noticed that vines, such as Virginia creeper, grape, and blackberry, continue to grow throughout the summer, long after trees stop putting on height. During intense competition with neighbors there is a race to get the leaves up, to get priority in grabbing the sunlight, and the leaves pop out and the twigs lengthen in a short sprint early in the summer. But the tree then has to route resources to thicken the trunk and limbs—the solid scaffolding to support the new leaves and twigs. I measured the girths of five different trees of different species throughout the season and indeed found that they did not increase in circumference until the leaves came on in June, and they stopped putting on girth by mid-July or August. (This conflicts with the idea of light and large and dark and narrow supposed summer versus winter ring growth. Is it summer versus *fall*?) Bushes such as honeysuckle put on as much length per season as trees such as oaks, but the first grows to be only ten feet tall whereas the other grows to 100 feet. The reason is that bushes put out shoots in all directions, and these shoots die off as fast as new ones are created; but trees grow in only one direction, and one year's growth adds to the next.

The geometry of trees is also an important aspect of leaf abscission. Deciduous trees spread their branches out and up in all directions to capture as much light as possible from above. However, long limbs, when snow-loaded, exert a huge torque that pulls them down until they may break or split the trunk. This configuration is great in the summer, for capturing lots of sunlight, but it compromises the ability to shed snow. In contrast, both spruces and firs have tough and generally short branches that bend down under a load, so that snow slides off. These trees are shaped like tents that partially collapse to the side but never split apart. The conical shape of these trees may also be an adaptation for capturing the sun's rays in the north, where the sunshine is much more lateral than it is farther south. Could leaf behavior also be a suitable alternative to being shed?

Animals' quick responses, though based on physiology, are called behavior. Some plants also have relatively fast responses that, though based on different mechanisms, are behavior as well. Many flowers track the sun so as to be continually warmed. A Venus flytrap leaf closes in seconds to capture an insect that lands on it. When touched, mimosa leaves droop and fold together in seconds, thus presenting less of a target for predators that might otherwise browse them. I was intrigued to learn that the leaves of rhododendrons in the mountains of northern China were curled into rolls during the winter (Schaller 2007). I could not believe my eyes when I saw the leaves of rhododendron of two species planted on our campus also rolling up. At temperatures below the freezing point of water (there are many days like that in Vermont!) the leaves drooped and rolled themselves into tight little tubes like sucking straws, and as soon as they were a couple of degrees warmer they again unrolled and raised themselves to the horizontal, looking as they do in spring and summer.

After being tightly rolled up (at minus 10°F), they could unroll and be almost totally flat and "normal" within about two to four minutes at room temperature (around 65°F). The reverse reaction occurred as well, but it was several minutes slower. I never did figure out what the mechanism underlying the curling is, but it has something to do with water, because when I cut twigs at room temperature and left their cut stems without access to water, the leaves slowly curled within about a day. When I then tried to rehydrate the curled leaves by wetting them and leaving them in either cold or warm water, they failed to respond in the winter, but did respond in the summer.

Ignoring for the time being the proximate question of the mechanism of how the leaf behaves, there is also the evolutionary question of why leaves roll up. Is there an adaptive advantage, and if so, what is it? Comparative data about the responses of other trees give clues.

One might suppose that the curling and drooping (rather than dropping) of leaves in the rhododendrons is a general response of leaves,

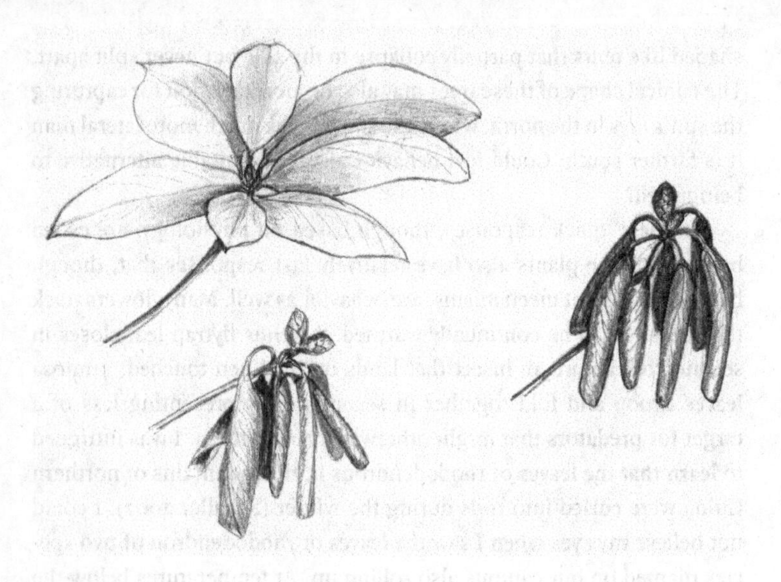

Fig. 39. **Leaves of rhododendrons drooping and curling in response to near-freezing temperatures.**

as such, to temperature. To find out, I raided some of our collection of house plants, taking a leaf from ten broad-leafed species and laying them outside at 0°F. A few minutes later I brought them back inside. As I had expected, all were frozen brittle-solid. A few minutes after thawing they were limp mush. None had curled. So the rhododendrons that survive at minus 30°F have a different physiology and a behavior that shows that frost-hardiness is possible in broad leaves. I called out-of-state relatives and had them send me twigs of broad-leafed magnolia from North Carolina and leatherleaf viburnum (*Viburnum rhytidophyllum*) from Pennsylvania (though viburnum is native to China). These plants' broad leaves, like those of the rhododendron, also survived the Vermont winter temperatures. In short, not only is it possible for broad leaves to stay alive in winter, but some broad leaves even behave in ways that greatly reduce their surface area at the temperatures where rain turns to snow.

The New England snowstorm of October 2005 had dumped a modest eight to twelve inches of snow in Vermont. We gauged its severity by the duration of the power outages, which in turn reflected the number of trees and branches that had fallen onto electric lines. (On our road, the power outage lasted three days. At other places throughout northern New England, it extended to more than a week.) Our costs were modest and temporary, but the lasting cost was to the trees caught with their leaves on. They paid either directly, in mortality, or in severe damage that would take decades to repair. Had the trees that shed their leaves later than others that summer made a large mistake by being off a few days in their timing, or on average is it a worse mistake to shed leaves too early?

Presumably, trees that are fully acclimatized to a generally stable climate shed their leaves at the optimum time, which is, by definition, when they get the best economic return over the long term. Their evolutionary calculation has balanced the appropriate amount of "insurance"—the cost of not gathering energy when the leaves are either shed early or delayed in being deployed, and the risk of the rare but severe injury. It seems that most of the time, many trees must shut down as though the summer has ended, long before it actually does end.

duration of the power outages, which in turn reflected the number of trees and branches that had fallen onto electric lines. On this road, the power outage lasted three days. At other places throughout northern New England, it extended to more than a week. Our costs were modest and temporary, but the lasting cost was to the trees caught with their leaves on; they paid either directly in mortality, or in severe damage that would take decades to repair. Had the trees that shed their leaves later than others that summer made a large mistake by being off a few days in their timing, or on average is it a worse mistake to shed leaves too early?

Presumably trees that are fully acclimatized to a generally stable climate shed their leaves at the optimum time, which is, by definition, when they get the best economic return over the long term. Their evolutionary calculation has balanced the appropriate amount of "insurance"—the cost of not gathering energy when the leaves are either shed early or delayed in being deployed, and the risk of the rare but severe injury. It seems that most of the time, many trees must shut down as though the summer has ended, long before it actually does end.

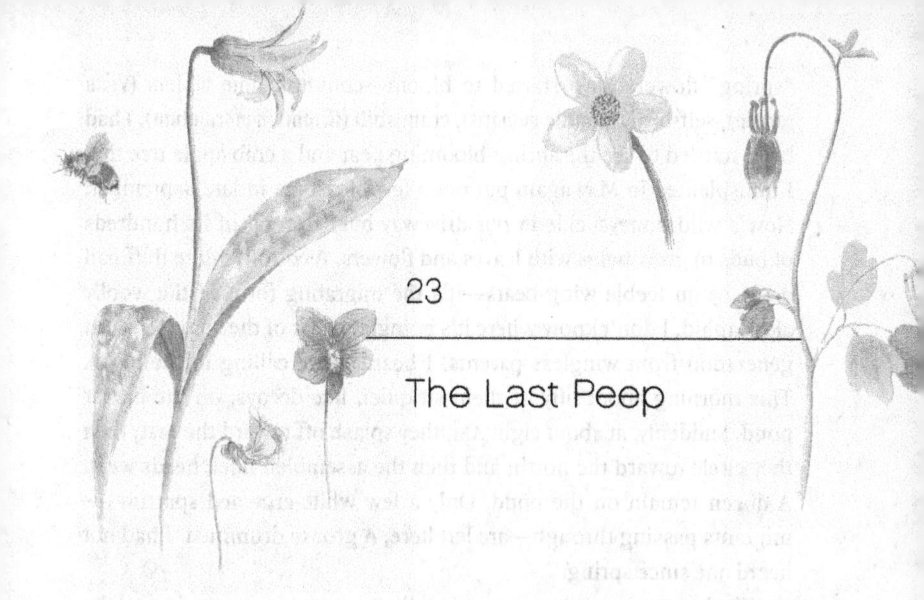

# 23

# The Last Peep

25 September 2005. IT IS A CRYSTAL-CLEAR DAY AND bees are bringing in pollen from the goldenrod, which is now fading fast. The purple and blue New England asters are still going strong, but the American ash are starting to shed their now purplish leaves. The sedges in the bog are brown and a few sugar maples are turning yellow. The first frost is forecast for tonight, so technically we're now beginning Indian summer. But we passed the fall equinox two days ago, when the sun spent twelve hours above and twelve hours below the horizon at every latitude on Earth. Here it is defined as the first day of fall (and it is the first day of spring in the southern hemisphere). From now on our days here will be gradually getting shorter than the nights, and that changing photoperiod will affect the physiology of trees, birds, and many mammals, to turn off growth and reproduction. It is therefore curious indeed that some spring-blooming plants are again showing signs of life—even our pear tree has a few blossoms. Dandelions are once again raising yellow flower heads. The phoebe sang briefly this morning after two months of silence, and spring peepers sometimes sound off with isolated calls at night.

19 October 2005. Windy, cool weather makes me feel restless and I'm looking forward to going to Maine tomorrow. The leaves are now falling thickly. Despite the cold and overcast weather a few more

"spring" flowers have started to bloom—common blue violets (*Viola sororia*), self-heal (*Prunella vulgaris*), cranesbill (*Geranium maculatum*). I had been startled to see the spring-blooming pear and a crab apple tree that I transplanted in May again put out a few blossoms in late September. Now a wild honeysuckle in our driveway burst several of its hundreds of buds to grow twigs with leaves and flowers. A cottony white fluff ball drifts by on feeble wing beats—it's the migrating form of the woolly alder aphid. I don't know where it's going, but it is of the summer's last generation from wingless parents. I heard geese calling in the night. This morning about fifty of them sit quiet, like decoys, on the beaver pond. Suddenly, at about eight AM, they splash off toward the east; then they circle toward the north; and then the assembled flock heads west. A dozen remain on the pond. Only a few white-crowned sparrows—migrants passing through—are left here. A grouse drummed. I had not heard one since spring.

The beavers in our bog are again felling poplar trees, gnawing off the branches, and dragging them into the water to store them in their food cache next to their lodge, where the ice will soon cover them. Chipmunks gather acorns and with their cheek pouches stuffed full they scurry to and from their dens to top off larders in their previously prepared underground storerooms. I stock the woodpile, harvest the honey, and insulate the house and hives, while Rachel obsessively cans vegetables and makes apple pies. Meanwhile, overhead the geese honk on their way south, and in the nearby woods the rutting of moose and deer is timed so that the young will be born early enough in the spring to grow up and withstand the next winter. As always when summer ends, most of what I see (and try to do) makes sense. It should. After all, few animals or plants would survive a full year without changing their behavior as well as their physiology to prepare for the awesome, inevitable winter challenge. Given the strict and predictable schedules of the seasons, I am puzzled by any plants or animals that would be fooled enough to be far off schedule. Are they aberrations? But if so, why are there so many of them?

I've just noticed that the raven pair living within earshot of our house have noisily returned to their nest site on the cliff, as though getting ready to renest. Invariably (so far) they will break off their apparent interest in a month or two (although there are reports of European ravens occasionally nesting in the fall). A friend reported seeing an osprey carrying a stick as though preparing to nest, and someone e-mailed to tell me he saw a raven pair near Bethel, Alaska, bringing sticks to their nest in mid-October. At this time of year the ruffed grouse occasionally drum in the woods, as they do in the spring when attracting mates. It has been claimed that this Indian summer activity is for "setting up territories," but instead most of the grouse are now almost semi-social, often feeding and resting together in small groups.

Woodpeckers also occasionally drum, and bluebirds have come and examined nest boxes. Other birds sing again, after a silence of at least two months. Daily I hear swamp, song, and white-throated sparrows; starlings; ruby-crowned kinglets; blue-headed vireos; and occasionally American robins, phoebes, and ovenbirds sound out abbreviated and somewhat hesitant renditions of their distinctive refrains, in what seems like a muted, halfhearted way. They usually give only the first few notes of their song at half volume and then it trails off as though they have changed their mind. (The spring migrants coming through do the same.)

Birdsong is a male prerogative that functions to lay claim to a territory and keep other males out, and possibly also to attract a mate. But many of these singing birds that I hear now are migrating south, moving through to their wintering grounds. None of them will form pair bonds or seek breeding territories until next spring and summer. In short, their singing is out of context and off schedule by about six months.

Perhaps singing is now a response of heightened exuberance that is normally reserved for the spring. But if so, it's a proximal, not an ultimate, response; exuberance can't explain plants' behavior. By late September I occasionally find not only the previously mentioned plants in flower in Vermont, but also at least one other species, the bunchberry, *Cornus canadensis*, in the Maine woods, near my camp. Bunchberry has flashy white flowers that carpet the north woods for two weeks in May. They are absent through the summer. When these flowers now reappear next to the plant's bright red berries and among fallen red, brown, and yellow tree leaves, they make a curious anomalous contrast.

None of the late flowers of the bunchberry will develop fruit. Many of them are misshapen as well, reminding me of the "imperfect," muted birdsong at this time. I suspect that unseasonably warm fall temperatures (global warming?) would cause even more flowers to bloom in the

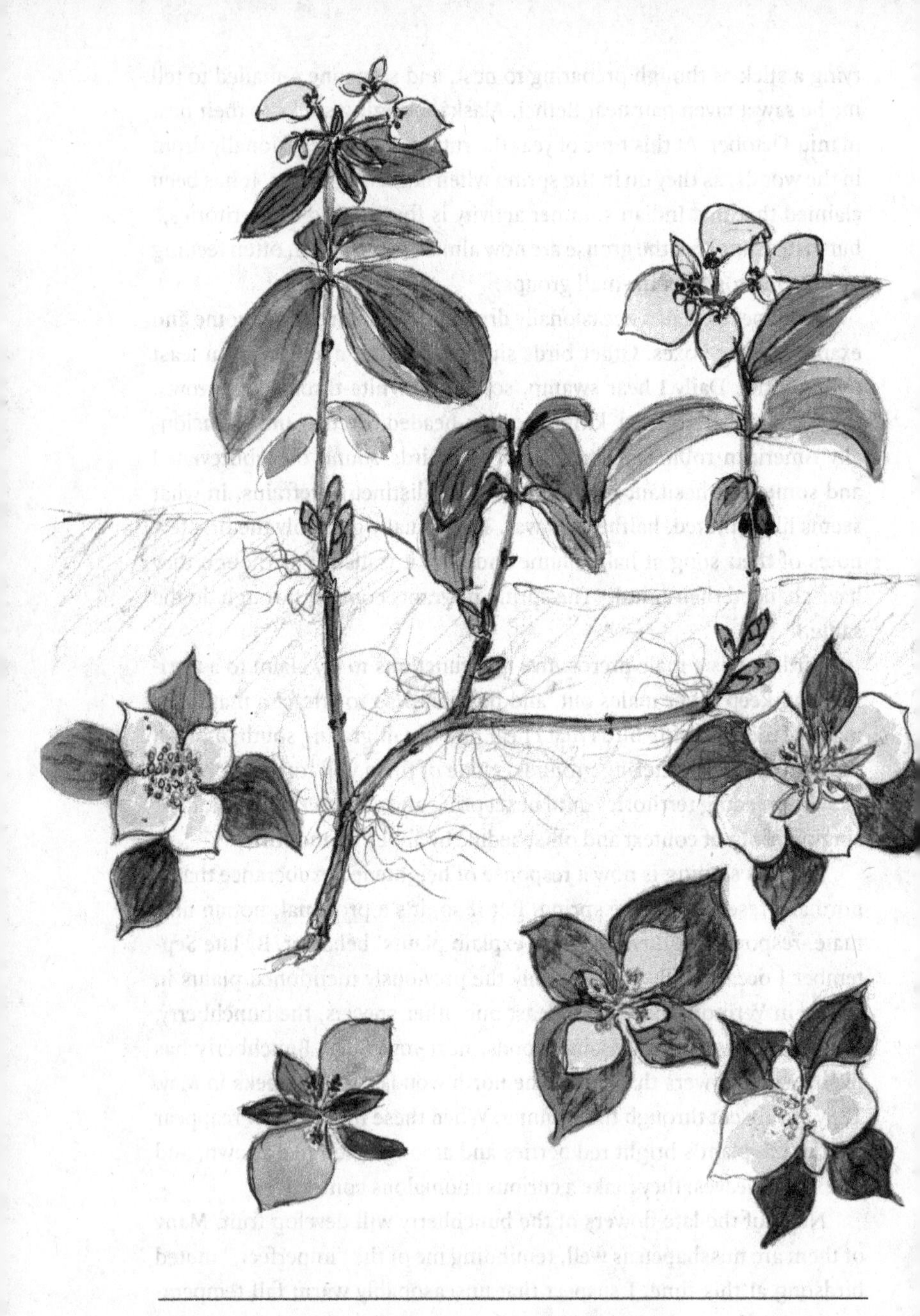

Fig. 40. **Many of the off-season bunchberry flowers are misshapen.**

fall, but temperature per se does not make them bloom, because it is invariably hotter in the previous months of summer, yet no flowering response is induced then.

Does the end of summer contain the seeds of spring? There are vast differences between summer and winter in temperatures and hours of light versus day, but there is much that is similar between the beginning and end of summer. Both tend to be cool times of year. And at the fall and spring equinoxes—22 September and 20 March, respectively—the photoperiod is identical: twelve to twelve (twelve hours of day and twelve of night).

Some flowers bloom in the spring, others in midsummer, and still others in the fall. In the laboratory one can induce one plant to flower under the influence of artificial short days, whereas another will flower only if subjected to long days. Similarly, a constant laboratory photoperiod induces birds to either lay eggs or shut off egg production. Wild birds breeding in the north start their reproductive cycles, including all their behavior of migration, courting, and nest building, in an orderly progression influenced by photoperiod. Our chickens lay throughout the summer when the natural photoperiod is on average at least thirteen hours of light and eleven of darkness. But to keep the hens laying in the middle of the winter all we have to do is keep a light on in their shed for several extra hours. At the spring equinox, near the end of March, the photoperiod is rapidly approaching thirteen to eleven, and that is also the time when many organisms are preparing for summer. Since they use the photoperiod to inform them of what season they are in, how does the same photoperiod, at around the fall equinox in late September and the spring equinox in late March, allow them to differentiate spring from fall?

Temperature is too variable to serve as a reliable cue to differentiate the beginning from the end of summer. Plants and animals not only need to know when summer is or is coming, but also need to predict when it will begin and end. Any given photoperiod as such is not the only answer. It seems remarkable enough that any organisms can measure photoperiod and almost universally respond appropriately to it, but they need an added mechanism to determine the direction of the changing photoperiod. That seems a lot to ask for. When plants flower at an inappropriate time, this effect is often attributed to "stress" or to unusually

Fig. 41. One bud (of eight on this section of twig) of a honeysuckle bush that has opened to extend a twig with leaves and flowers in October; normally the plant blooms in late May.

high temperatures. Stress could indeed be a contributing factor, but perhaps the springtime photoperiod in the fall is in itself a stressor.

Mistakes or imperfections provide variety for natural selection to work on, to permit evolution. There is even a mechanism whose only "purpose" is to produce variation. It's called sex. If there were no variation, there could be no evolution. If species had been magically created, they would all be clones. It was a heritable mutation that caused central European blackcap warblers to misorient, so that they ended up flying west in the fall rather than south to Africa. These birds got a new population started, and they are now thriving in Great Britain. Some possibly misguided phoebes who abandoned cliffs and started nesting on houses became the norm because houses were safer. Salmon that did not find their home streams on their spawning migration and by chance ended up in other streams eventually colonized the new streams, expanding the populations. Possibly confused woodpeckers that wasted a lot of time and energy hammering pseudo nest holes in the fall found them to be useful places for overnighting during very cold weather, and they had a slightly better chance of surviving than those who seemed to be less misguided. Similarly, the rare blue violet that blooms in the fall reminds me that nature is not always proximally perfect, though it evolves and ultimately persists because of its imperfections.

FOR YEARS ON WARM NIGHTS AND SOMETIMES ALSO ON warm days in late summer and fall, I have heard strange, generally isolated high-pitched cheeps and chirps coming from our woods. Whenever I have stalked near to determine the sources of these birdlike calls, they have invariably stopped and I have seen nothing. It has taken me a while to finally establish that I was hearing the voices of apparently misguided spring peepers and wood frogs. In the spring these frogs collectively make a deafening din in their breeding ponds, and afterward they hop back onto land in the nearby woods, where they remain silent all summer. However, by September and early October, in Indian summer, when you again begin to hear their voices, they are never in breeding ponds but always in the woods, where they will overwinter under the fallen leaves. No bullfrog, leopard frog, or green frog makes the same "mistake" of calling as though jumping the gun to begin the breeding ritual six months ahead of schedule. But wood frogs do call, although not as frequently as the peepers and although the calls are usually very brief and isolated. Once while I was up in a spruce tree in Maine during November I suddenly heard a wood frog below me; then another joined it; then a third. They were separated by about 100 feet. There was no pool in sight. The three called for about ten minutes, and then resumed their silence. These were probably their last peeps of the season.

These fall-calling frogs are already filled with egg masses like those they will deposit half a year later, as I found out by accident. I had built an aviary enclosing a section of woods, and one morning at dawn in September one of the ravens inside captured and killed a frog. I immediately confiscated it to make a positive identification: it was a plump female wood frog carrying a full cluster of eggs that looked identical to what females freshly out of hibernation deposit in their breeding pools in April. If the frog had frozen the next day and thawed out in April, it would have awakened to another day of very similar temperature and photoperiod, and it might not know the difference, or might not know that anything had changed. The seven-month interval until these eggs would have been laid would have been, to a cold or frozen frog under leaves and snow and ice, a time of death when a minute is an eternity and an eternity a minute. The end of summer is also the beginning.

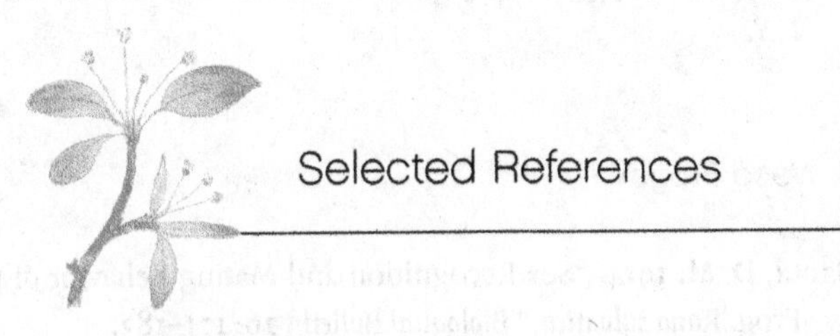

# Selected References

## 1. Preparing for Summer


Bünning, E. 1973. *The Physiological Clock*, 3rd ed. Springer Verlag, Berlin, Heidelberg.

Denny, F. E., and E. N. Stanton. 1928. "Localization of Response of Woody Tissue to Chemical Treatment That Breaks the Rest Period," *American Journal of Botany* 15: 337–344.

Gwinner, E., and W. Wiltschko. 1980. "Circannual Changes in Migratory Orientation in the Garden Warbler, *Sylvia borin*," *Behavioral Ecology and Sociobiology* 7: 73–78.

Gwinner, E. 1986. *Zoophysiology*, Vol. 18, *Circannual Rhythms*. Springer Verlag, Berlin, Heidelberg, New York.

Pengelley, E. T., R. C. Aloia, and B. Barnes. 1979. "Circannual Rhythmicity in the Hibernating Ground Squirrel, *Citellus lateralis*, under Constant and Hyperthermic Ambient Temperature," *Comparative Biochemistry and Physiology* 61A: 599–603.

Schmidt-Koenig, K. 1975. *Migration and Homing in Animals*. Springer Verlag, Berlin, Heidelberg, New York.

## 2. Awakening

Pearson, T. G. 1917. *Birds of America*. Garden City Books, Garden City, N.Y.

## 3. Wood Frogs

Banta, D. M. 1914. "Sex Recognition and Mating Behavior of the Wood Frog, *Rana sylvatica*," *Biological Bulletin* 26: 171–183.

Berven, K. A. 1981. "Mate Choice in the Wood Frog, *Rana sylvatica*," *Evolution* 35: 707–722.

Berven, K. A., and T. A. Grudzen. 1991. "Dispersal in the Wood Frog (*Rana sylvatica*): Implications for Genetic Population Structure," *Evolution* 44: 2047–2056.

Cornell, T. J., K. A. Berven, and G. J. Gamboa. 1989. "Kin Recognition by Tadpoles and Froglets of the Wood Frog *Rana sylvatica*," *Oecologia* 78: 312–316.

Emlen, S. T. 1968. "Territoriality in the Bullfrog, *Rana catesbeiana*," *Copeia* 1968: 240–243.

———. 1976. "Lek Organization and Mating Strategies of the Bullfrog," *Behavioral Ecology and Sociobiology* 1: 283–313.

Emlen, S. T., and L. W. Oring. 1977. "Ecology, Sexual Selection, and the Evolution of Mating Systems," *Science* 197: 215–223.

Howard, R. D. 1978. "The Evolution of Mating Strategies in Bullfrogs, *Rana catesbeiana*," *Evolution* 32: 859–871.

———. 1980. "Mating Behaviour and Mating Success in Woodfrogs, *Rana sylvatica*," *Animal Behaviour* 28: 705–616.

Howard, R. D., and A. G. Kluge. 1985. "Proximate Mechanisms of Sexual Selection in Wood Frogs," *Evolution* 39: 260–277.

Taigen, T. L., and K. D. Wells. 1985. "Energetics of Vocalization by an Anuran Amphibian (*Hyla versicolor*)," *Journal of Comparative Physiology* 155: 163–170.

Waldman, B. 1982. "Adaptive Significance of Communal Oviposition in the Wood Frog (*Rana sylvatica*)," *Behavioral Ecology and Sociobiology* 10: 169–174.

Wells, K. D. 1977. "The Social Behaviour of Anuran Amphibians," *Animal Behaviour* 25: 666–693.

———. 1977. "Territoriality and Mating Success in the Green Frog (*Rana clamitans*)," *Ecology* 58: 750–762.

5. Bald-Faced Hornet Nests

Heinrich, B. 1984. "Strategies of Thermoregulation and Foraging in Two Wasps, *Dolichovespula maculata* and *Vespula vulgaris*," *Journal of Comparative Physiology B* 154: 175–180.

Levi, P. 1984. *The Period Table*. Schocken, New York.

6. Mud Daubers and Behavior

Blackledge, T. A., and K. M. Pickett. 2000. "Predatory Interactions between Mud-Dauber Wasps (Hymenoptera, Sphecidae) and Argiope (Aranaeae, Araneidae) in Captivity," *Journal of Arachnology* 28: 211–216.

Brockmann, H. J. 1980. "Diversity in the Nesting Behavior of Mud-Daubers (*Trypoxylon politum* Say: Sphecidae)," *Florida Entomologist* 63: 53–64.

Heinrich, B., and T. M. Casey. 1978. "Heat Transfer in Dragonflies: 'Flyers' and 'Perchers,'" *Journal of Experimental Biology* 74: 17–36.

Muma, M. H., and W. F. Jeffers. 1945. "Studies of the Prey of Several Mud-Dauber Wasps," *Annals of the Entomological Society of America* 38: 245–257.

O'Neil, K. M., J. F. O'Neil, and R. P. O'Neil. 2007. "Sublethal Effect of Brood Parasitism on the Grass-Carrying Wasp, *Isodontia mexicana*," *Ecological Entomology* 32(1): 123–127.

Shafer, G. D. 1949. *The Ways of a Mud Dauber*. Stanford University Press, Palo Alto, Calif.

7. The Blues

Als, T. D., R. Vila, N. P. Kandul, et al. 2004. "The Evolution of Alternative Parasitic Life Histories in Large Blue Butterflies," *Nature* 432: 386–390.

Bingham, C. T. 1907. "Fauna of British India," *Butterflies* 2.

Braby, M. F. 2000. *Butterflies of Australia*. Melbourne, CSIRO, Vol. 2, pp. 622–623.

Eastwood, R., and A. J. King. 1998. "Observations on the Biology of *Arhopala wildei* Miskin (Lepidoptera: Lycaenidae) and Its Host Ant *Polyrhachis queenslandica* Emery (Hymenoptera: Formicidae)," *Australian Entomologist* 25(1): 1–6.

Lohman, D. J., Q. Liao, and N. E. Pierce. 2006. "Convergence of Chemical Mimicry in a Guild of Aphid Predators," *Ecological Entomology* 31: 41–51.

8. Artful Diners

Dussourd, D. E., and R. F. Denno. 1991. "Deactivation of Plant Defenses: Correspondence between Insect Behavior and Secretory Canal Architecture," *Ecology* 72: 1383–1396.

Greene, E. 1989. "A Diet-Induced Developmental Polymorphism in a Caterpillar," *Science* 243: 643–646.

Heinrich, B. 1971. "The Effect of Leaf Geometry on the Feeding Behaviour of the Caterpillar of *Manduca sexta* (Sphingidae)," *Animal Behaviour* 19: 119–124.

———. 1979. "Foraging Strategies of Caterpillars: Leaf Damage and Possible Predator Avoidance Strategies," *Oecologia* 40: 325–337.

———. 1980. "Artful Diners," *Natural History* 86: 42–51.

Heinrich, B., and S. L. Collins. 1983. "Caterpillar Leaf Damage and the Game of Hide-and-Seek with Birds," *Ecology* 64: 592–602.

Holmes, R. T., J. C. Schultz, and P. Nothnagle. 1979. "Bird Predation on Forest Insects: An Exclosure Experiment," *Science* 206: 462–463.

Real, P. G., R. Ianazzi, A. C. Kamil, and B. Heinrich. 1984. "Discrimination and Generalization of Leaf Damage by Blue Jays (*Cyanocitta cristata*)," *Animal Learning and Behavior* 12: 202–208.

Salazar, B. A., and D. W. Whitman. 2001. "Defensive Tactics of Caterpillars against Predators and Parasitoids." In T. N. Ananthakrishnan (ed.), *Insects and Plant Defence Dynamics*, pp. 101–207. Science Publishers, Enfield, N.H.

Stamp, N. E., and T. M. Casey (eds.). 1993. *Caterpillars: Ecological and Evolutionary Constraints on Foraging*. Chapman and Hall, New York.

Wagner, David, L. 2005. *Caterpillars of Eastern North America*. Princeton University Press, Princeton, N.J.

## 9. Masters of Disguise

Aiello, A., and R. E. Silberglied. 1978. "Life History of *Dynastor darius* (Lepidoptera: Nymphalidae: Brassolinae) in Panama," *Psyche* 85: 331–345.

Booth, F. W., and P. D. Neufer. 2005. "Exercise Controls Gene Expression," *American Scientist* 93: 28–35.

Collins, M. M. 1999. "A Hostplant-Induced Larval Polymorphism in *Hyalophora euryalus* (Saturniidae)," *Journal of the Lepidopterists' Society* 53: 22–28.

Fink, L. S. 1995. "Foodplant Effects on Colour Morphs of *Eumorpha fasciata* Caterpillars (Lepidoptera: Sphingidae)," *Biological Journal of the Linnaean Society* 56: 423–437.

Gibb, J. A. 1962. "L. Tinbergen's Hypothesis of the Role of Specific Search Images," *Ibis* 104: 106–111.

Grayson, J., and M. Edmunds. 1989. "The Causes of Color and Color-Change in Caterpillars of the Poplar Hawkmoths (*Laothoe populi* and *Smerinthus ocellata*)," *Biological Journal of the Linnaean Society* 37: 263–279.

Greene, E. 1989. "A Diet-Induced Developmental Polymorphism in a Caterpillar," *Science* 243: 643–646.

Hudson, A. 1966. "Protein in the Haemolymph and Other Tissues of the Developing Tomato Hornworm, *Protoparce quinquemaculata* Haworth," *Canadian Journal of Zoology* 44: 541–555.

Miller, J. C., D. H. Janzen, and W. Hallwachs. 2006. *Caterpillars: Portraits from the Tropical Forests of Costa Rica*. Harvard University Press, Cambridge, Mass., and London.

Schneider, G. 1973. "Ueber den Einfluss verschiedener Umweltfaktoren

auf den Faerbungspolyphaenismus der Raupen des tropisch-amerikanischen Schwaermers *Erinnyis ello* L. (Lepidopt. Sphingid.)," *Oecologia* (Berlin) 11: 351–370.
Suzuki, Y., and H. Frederik Nijhout. 2006. "Evolution of a Polymorphism by Genetic Accommodation," *Science* 311: 650–651.
Truman, J. W., L. M. Riddiford, and L. Safranek. 1973. "Hormonal Control of Cuticle Coloration in *Manduca sexta*: Basis of an Ultrasensitive Bioassay for Juvenile Hormone," *Journal of Insect Physiology* 19: 195–203.
Wagner, D. L. 2005. *Caterpillars of Eastern North America*. Princeton University Press, Princeton, N.J.

## 10. Cecropia Moths

Beckage, N. E. 1997. "The Parasitic Wasp's Secret Weapon," *Scientific American* (November): 82–87.
Cruz, Y. P. 1981. "A Sterile Defender Morph in a Polyembryonic Hymenopteran Parasite," *Nature* 294: 446–447.
Marsh, F. L. 1937. "Ecological Observations upon Enemies of Cecropia, with Particular Reference to Its Hymenopterous Parasites," *Ecology* 18: 106–112.
———. 1941. "A Few Life-History Details of *Samia cecropia* within the Southwestern Limits of Chicago," *Ecology* 22: 331–337.

## 11. Calosamia Collapse

Benson, J., A. Pasquele, R. G. Van Driesche, and J. Elkinton. 2003. "Introduced Braconid Parasitoids and Range Reduction of a Native Butterfly in New England," *Biological Control* 28: 197–213.
Boettner, G. H., J. S. Elkinton, and C. J. Boettner. 2000. "Effects of a Biological Control Introduction on Three Nontarget Native Species of Saturniid Moths," *Conservation Biology* 14: 1798–1806.
Elkinton, J. S., D. Parry, and G. H. Boettner. 2006. "Implicating an Introduced Generalist Parasitoid in the Invasive Browntail Moth's Enigmatic Demise," *Ecology* 87: 2664–2672.

Godfray, H. C. J. 1995. "Field Experiments with Genetically Manipulated Insect Viruses: Ecological Issues," *Trends in Ecology and Evolution* 10: 465–469.

Marshall, S. A. 2006. *Insects: Their Natural History and Diversity*. Firefly, Buffalo, N.Y.

Peigler, R. S. 1994. "Catalogue of Parasitoids of Saturniidae of the World," *Journal of Research on the Lepidoptera* 33: 1–121.

Smith, M. A., D. M. Wood, D. H. Janzen, et al. 2007. "DNA Barcodes Affirm That 16 Species of Apparently Generalist Parasitoid Flies (Diptera, Tachinidae) Are Not All Generalists," *Proceedings of the National Academy of Sciences* 104: 4967–4972.

12. New England Longhorns

Brisley, R. B., and R. A. Channel. 1924. "The Oak Girdler, *Oncideres quercus* Skinner," *Journal of Economic Entomology* 17: 159.

Linsley, E. G. 1940. "Notes on *Oncideres* Twig Girdlers," *Journal of Economic Entomology* 33: 561–563.

———. 1959. "Ecology of the Cerambycidae," *Annual Review of Entomology* 4: 99–138.

Linsley, E. G., and J. A. Chemsak. 1961–1995. *The Cerambycidae of North America*. (9 Parts.) University of California Publications in Entomology. (See 18:1–135; 19:1–102; 20: 1–188; 21: 1–165; 22: 1–197; 69:1–138; 80:1–186; 102:1–258; 114:1–292.)

Stride, G. O., and E. P. Warwick. 1962. "Ovipositional Girdling in a North American Cerambycid Beetle, *Mecas saturnine*," *Animal Behaviour* 10: 112–117.

Yanega, D. 1996. *Field Guide to Northeastern Longhorn Beetles* (Coleoptera: Cerambycidae). Manual 6. Illinois Natural History Survey, Champaign.

14. The Hummingbird and the Woodpecker

Bent, A. C. 1992. *Life Histories of North American Woodpeckers*. Indiana University Press, Bloomington and Indianapolis.

Grantsau, R. 1988. *Die Kolibris Brasiliens.* Expressao E Cultura, Rio de Janeiro, Brazil.

Kilham, L. 1983. *Life History Studies of Woodpeckers of Eastern North America.* Nuttall Ornithological Club, No. 20. Cambridge, Mass.

———. 1992. *Life History Studies of Eastern North America.* Dover Publications, New York..

Robinson, T. R., R. R. Sargent, and M. B. Sargent. 1996. "Ruby-Throated Hummingbirds (*Archilochus colubris*)." In A. Poole and F. Gill (eds.), *The Birds of North America,* No. 204. American Ornithologists' Union, Washington, D.C.

Sargent, R. 1999. *Wild Bird Guides: Ruby-Throated Hummingbird.* Stackpole, Mechanicsburg, Pa.

Stichter, S. 2004. "Fall 2003 Migration of Ruby-Throated Hummingbirds in New England," *Bird Observer* 32(1): 12–23.

Walters, E. L., and P. E. Lowther. "Yellow-Bellied Sapsucker, *Sphyrapicus varius.*" In A. Poole and F. Gill (eds.), *The Birds of North America,* No. 662. American Ornithologists' Union, Washington, D.C.

Willimont, L. A., S. E. Senner, and L. J. Goodrich. 1988. "Fall Migration of Ruby-Throated Hummingbirds in the Northeastern United States," *Wilson Bulletin* 100(3): 482–488.

## 16. Extreme Summer

Adolf, E. F. 1947. *Physiology of Man in the Desert.*

Alcock, John. 1990. *Sonoran Desert Summer.* University of Arizona Press, Tucson.

Blagden, C. 1775. *Further Experiments and Observations in a Heated Room.* Phil. Trans. Roy. Soc. London 65 (1): 111–123

Brain, C. K. 1981. *Hunters or the Hunted? An Introduction to African Cave Taphonomy.* University of Chicago Press, Chicago and London.

Brown, G. W., Jr. 1968. *Desert Biology,* Vol. I. Academic, New York and London.

Conway, J. R. 2008. "Sweet Dreams," *Natural History* 117(3): 30–35.

Cowles, Raymond. 1977. *Desert Journal: Reflections of a Naturalist.* University of California Press, Berkeley and Los Angeles.

Evenari, M., L. Shanan, and N. Tadmor. 1982. *The Negev: The Challenge of a Desert*. Harvard University Press, Cambridge, Mass., and London.

Heath, J. E., and P. J. Wilkin. 1970. "Temperature Responses of the Desert Cicada: *Diceroprocta apache* (Homoptera. Cicadidae)," *Physiological Zoology* 43: 145–154.

Heinrich, B. 1984. "Strategies of Thermoregulation and Foraging in Two Wasps, *Dolichovespula maculata* and *Vespula vulgaris*," *Journal of Comparative Physiology* B 154: 175–180.

———. 1996. *The Thermal Warriors: Strategies of Insect Survival*. Harvard University Press, Cambridge, Mass., and London.

Lee, D. H. K. 1968. "Human Adaptation to Arid Environments." In G. W. Brown Jr., *Desert Biology*, Vol. I. Academic, New York and London, pp. 517–563.

Martin, H. 1958. *The Sheltering Desert*. Thomas Nelson, Edinburgh, New York, and Toronto.

Ono, M., I. Igarashi, E. Ohno, and M. Sasaki. 1995. "Unusual Thermal Defense by a Honeybee against a Mass Attack by Hornets," *Nature* 377: 334–336.

Shkolnik, A. 1982. "Adaptations of Animals to Desert Conditions." In M. Evenari, L. Shanan, and N. Tadmor, *The Negev: The Challenge of a Desert*, 2nd ed. Harvard University Press, Cambridge, Mass., pp. 301–323.

Thomas, E. M. 1958/1989. *The Harmless People*, 2nd ed. Vintage, New York.

Toolson, E. C. 1987. "Water Profligacy as an Adaptation to Hot Deserts: Water Loss Rates and Evaporative Cooling in the Sonoran Desert Cicada, *Diceroprocta apache* (Homoptera: Cicadidae)," *Physiological Zoology* 60: 379–385.

Walsberg, G. E. 1982. "Coat Color, Solar Heat Gain, and Conspicuousness in Phainopepla," *Auk* 99: 495–502.

Wehner, R. A. 1976. "Polarized Light Navigation by Insects," *Scientific American* (July).

Wehner, R. A., C. Marsh, and S. Wehner. 1992. "Desert Ants on a Thermal Tightrope," *Nature* 357: 586–587.

## 17. Moss, Lichens, and Tweedlaarkanniedood

Bornman, C. H. 1978. *Paradox of a Parched Paradise*. C. Struik, Cape Town and Johannesburg.

Cocchietto, M., N. Skert, and P. L. Nimis. 2002. "A Review on Usnic Acid, an Interesting Natural Compound," *Naturwissenschaften* 89: 137–146.

Henderson, L. L. 1913. *The Fitness of the Environment: An Inquiry into the Biological Significance of the Properties of Matter*. Macmillan, New York.

Lawrey, J. D. 1989. "Lichen Secondary Compounds: Evidence for a Correspondence between Anti Herbivory and Anti Microbial Function," *Bryologist* 92: 326–328.

## 18. Perpetual Summer Species

Bramble, D. M., and D. E. Lieberman. 2004. "Endurance Running and the Evolution of *Homo*," *Nature* 432: 345–352.

Chen, I. 2006. "Born to Run," *Discover* (May).

Heinrich, B. 2001. "Endurance Predator," *Outside* 25(9): 70–76.

Hoffecker, J. F. 2005. *A Prehistory of the North*. Rutgers University Press, New Brunswick, N.J.

Liebenberg, L. 2006. "Persistence Hunting by Modern Hunter-Gatherers," *Current Anthropology* 47: 6.

Lieberman, D. L., D. M. Bramble, D. A. Raichler, and J. J. Shea. 2007. "The Evolution of Endurance Running and the Tyranny of Ethnography: A Reply to Pickering and Bunn (2007)," *Journal of Human Evolution* 53(4): 439–442.

Pickering, T. R., and H. T. Bunn. 2007. "The Endurance Running Hypothesis and Hunting and Scavenging in Savanna-Woodland," *Journal of Human Evolution* 53(4): 434–438.

Rantala, M. J. R. 2007, "Evolution of Nakedness in *Homo sapiens*," *Journal of Zoology* 273: 1–7.

Stringer, C., and C. Gamble. 1993. *In Search of the Neanderthals*. Thames and Hudson, New York.

Trinkaus, E., and P. Shipman. 1993. *The Neanderthals*. Knopf, New York.

Weyand, P. G., and J. A. Davis. 2005. "Running Performance Has a Structural Basis," *Journal of Experimental Biology* 208: 2625–2631.

19. Ant Wars

Heinrich, B., and M. E. Heinrich. 1984. "The Pit-Trapping Strategy of the Ant Lion, *Myrmeleon immaculatus* DeGeer (Neuroptera: Myrmeleontidae)," *Behavioral Ecology and Sociobiology* 14: 141–160.

Wilson, E. O. 1990. *The Ants*. Harvard University Press, Cambridge, Mass.

22. Ending Summer

Fuente, R. K., and A. C. Leopold. 1968. "Senescence Processes in Leaf Abscission," *Plant Physiology* 43(9 Part B): 1496–1505.

Heinrich, B. 1996. "When the Bough Bends," *Natural History* 2(96): 56–57.

———. 1997. "Construction for Strength," *Vermont Woodlands* (Winter): 16–19.

———. 2001. "O Tannenbaum," *Natural History* 110(10) 38–39.

———. 2006. "Timing Is Everything," *Northern Woodlands* (Autumn): 46–48, 75–80.

Schaller, G. B. 2007. *A Naturalist and Other Beasts: Tales from a Life in the Field*. Sierra Club Books, San Francisco, Calif.

23. The Last Peep

Hauri, R. 1968. "Horstbau beim Kolkraben im Herbst," *Ornithologischer Beobachter* 65: 28–29.

Hyatt, J. H. 1946. "Ravens Nest in October," *British Birds* 39: 83–84.

Mearns, R., and B. Mearns. 1989. "Sucessful Autumn Nesting of Ravens," *Scottish Birds* 15:179.

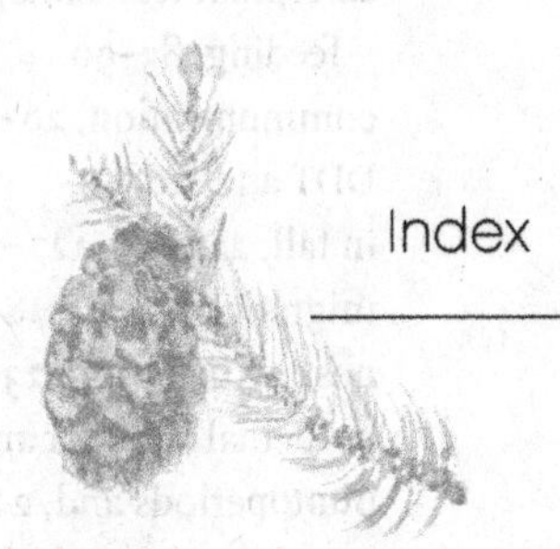

# Index

## I



## J



## K



## L



## M

N



O



P

Q



R



S

T

## X



## Y

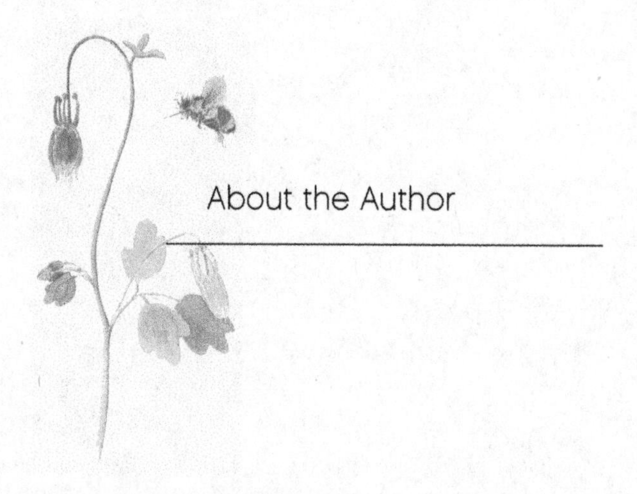

## About the Author

Bernd Heinrich is the author of numerous award-winning books, including the best-selling *Winter World*, *Mind of the Raven*, and *Why We Run*, and has received countless honors for his scientific work. He also writes for *Scientific American*, *Outside*, *American Scientist*, and *Audubon*, and has written book reviews and op-eds for the *New York Times* and the *Los Angeles Times*. He studied at the University of Maine and UCLA, and is professor emeritus of biology at the University of Vermont. Heinrich divides his time between Vermont and the forests of western Maine.

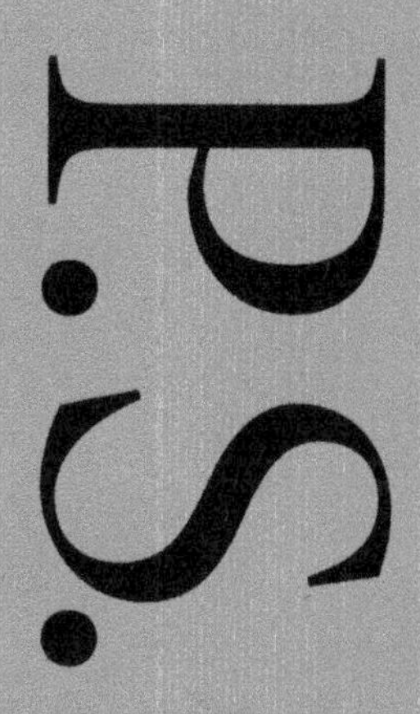

Insights,
Interviews
& More . . .

## About the author



## About the book



## Read on

# Meet Bernd Heinrich

© 2007 RACHEL SMOLKER

BERND HEINRICH is the author of numerous books, including *The Trees in My Forest*, *A Year in the Maine Woods*, *Ravens in Winter*, *One Man's Owl*, *Bumblebee Economics* (nominated for the National Book Award), and *Mind of the Raven* (winner of the John Burroughs Medal for Natural History Writing). He is professor emeritus of biology at the University of Vermont. Heinrich divides his time between Vermont and the forests of western Maine.

# A World of Permanent Summer?

WHEN I WROTE *SUMMER WORLD* I wanted to highlight the differences between the worlds of summer and winter. But the real world that we live in is a biological one, where plants and animals inhabit tightly knit communities in life cycles that are locked into annual as well as daily rhythms that make a whole. What, we might wonder, would the world be like if a continuous summer (or winter) were thrust upon them in a mere instant of a century or two, relative to the millions of years it took the organisms to adjust by evolution?

Until very recently the question would have been theoretical. Not any more. Glaciers are melting all over the world. We are experiencing altered seasons that impact our finely tuned nature in unimaginable ways. We all know what is happening, even though here in mid-latitudes we will not experience global warming like the Inuits of the far North, or the islanders in the vast Pacific, or the millions of starving people in Africa and Asia who are becoming displaced refugees from drought. Our science is getting better, allowing us to predict the future, and it looks ominous. Acidification of the oceans and the collapse of the marine food chain can be predicted from reasonable data and assumptions. But I do not wish to here engage in the endless debates about climate and policies. These are by now well known and publicized. I cite only some signs of effects on several local little creatures.

Some ten or twenty years ago ▸

About the book

“What, we might wonder, would the world be like if a continuous summer (or winter) were thrust upon them in a mere instant of a century or two, relative to the millions of years it took the organisms to adjust by evolution?”

I happened to read a technical report in an obscure journal that mentioned a huge die-off of carabid beetles. Most people never see these flightless predatory ground beetles because they do not look for them and would not recognize one if they saw it. I happened to be interested because I had collected them as a boy, and my carabid collection was my pride and joy. It had provided me with endless hours in the fields and woods. I was also interested in the paper because the author had studied the cause of the die-off, and he concluded that it was the result of a mild winter. I was surprised at this, because I had expected a severe winter to be worse.

A slightly unusual higher temperature had upset the beetles' energy balance. Like almost all hibernators, these beetles do not feed for six months, relying on fat stored in their bodies. The elevated temperature had sped up their metabolism so that they ran out of fat before food was available in the spring. The same phenomenon affects hibernating bats. Additionally, hibernating insects are strongly affected by the pattern of temperature; a fluctuating temperature may switch the animal back and forth from winter to summer survival mode, and can cause death due to insufficient time to adjust. It is as dangerous for them to be too warm as it is too cold, particularly if that cold is interspersed with warm spells. Insect physiologists have also discovered that some insects will not molt from pupa to adult until the pupa has experienced enough cold to allow it to "know" that winter has occurred and it is safe to become an adult that needs food and may live only days.

I thought little about global warming then. The idea seemed almost preposterous to me. But I was chilled when I read about these obscure beetles, because as many insects in northern climates have similar life cycles, they should be equally vulnerable. And while most carabids are predators of caterpillars and other small invertebrate animals, they might not be missed. But the hundreds of species of bees that hibernate in the winter and pollinate flowers in the summer are needed to produce seeds and are hence vital for plant reproduction. They are keystone organisms of the ecosystem. Since the 1970s several very common species of bumblebees have become virtually extinct in northern America. We know it is not because of pesticides. What is it, then? Can it be some kind of climate disruption that affects their overwintering? Many hundreds of species of little wasps also perform a keystone function, and they are even less known that almost any beetles or bees. These wasps parasitize the thousands of species of moth caterpillars that all eat leaves. Without those wasps (many of whom are almost microscopic) in an ecosystem, "instant" worldwide defoliation of forests can be almost guaranteed. And we don't have the slightest idea if warmer winter temperatures will inhibit their annual survival or enhance it.

Each species has, through millions of years, adapted to a specific temperature regime, and any change will have an effect. In northern Alaska, after a series of warm ▸

"Since the 1970s several very common species of bumblebees have become virtually extinct in northern America. We know it is not because of pesticides. What is it, then?"

**A World of Permanent Summer?** ***(continued)***

summers, the spruce budworm moth appeared and seriously weakened trees. The spruce bark beetle then took over and underwent a population explosion. Nearly three million acres of coniferous forests were decimated. Fires from lightning strikes consumed the dead, dry forests, and millions of charred acres both in Alaska and the western United States are witness. The beetle had always been there, but warmer summers allowed it to complete its life cycle in one year rather than the normal two, and warmer winters produced less mortality. A slightly higher reproductive rate then boosted the population, so hordes could attack and multiply even faster, overpowering the trees' defenses that had been safe before.

I did not think global warming could affect us here in New England. We tend to dread the cold winters, and when there is talk about global warming many people say, "Bring it on!" One thing is certain: it will come. And "diseases" of the ecosystem may be the first signs.

Several years ago I saw a bull moose that had recently keeled over in the snow in the Maine woods near my cabin. The early April snow all around him was black and crawling with winter, or "moose," ticks. These individuals—all adults—would now die with him. But they normally drop off by late April (moose are free of these ticks in the summer). If they fall onto snow, they die; but if instead the ground is snow-free, they live, and each female then lays from 3,000 to 4,000 eggs before she dies. The nymphs hatching from the eggs hitch a

"I did not think global warming could affect us here in New England. We tend to dread the cold winters, and when there is talk about global warming many people say, 'Bring it on!'"

ride onto another moose in the fall. But if the fall is cold and wet, they don't catch a ride. Thus if there is a short winter, these winter ticks increase their survival chances at both the fall and spring ends of their annual life cycle.

One surviving female tick has a theoretical capacity of having four million grand-ticks (if they lived). Of course not all of them survive—not even close. But nevertheless, infected moose can now sometimes get up to fifty ticks per square inch of hide. The winter ticks cause not only anemia due to blood loss but itching. Afflicted moose rub their fur off in their futile attempts to rid themselves of the parasites, and they become weakened due to increased heat loss and less feeding from the distraction. Then they die.

Most tick species require summer warmth, and warmer summers breed more ticks. The deer ticks that spread Lyme disease are not yet able to live in most parts of Maine, but they are spreading north. I cannot imagine the Maine woods without moose, or Vermont without maple trees. But I can imagine the impact when we can't walk in the woods for fear of disease infection. Will we then become even more oblivious and indifferent to the signs of nature?

# Have You Read?

**THE SNORING BIRD: MY FAMILY'S JOURNEY THROUGH A CENTURY OF BIOLOGY**

From Bernd Heinrich comes the remarkable story of his father's life, his family's past, and how the forces of history and nature have shaped his own life. Although Bernd Heinrich's father, Gerd, a devoted naturalist, specialized in wasps, Bernd tried to distance himself from his "old-fashioned" father, becoming a hybrid: a modern, experimental biologist with a naturalist's sensibilities.

**WINTER WORLD: THE INGENUITY OF ANIMAL SURVIVAL**

From flying squirrels to grizzly bears, from torpid turtles to insects with antifreeze, the animal kingdom relies on some staggering evolutionary innovations to survive winter. Unlike their human counterparts, who must alter the environment to accommodate physical limitations, animals are adaptable to an amazing range of conditions.

Examining everything from food sources in the extremely barren winter landscape to the chemical composition that allows certain creatures to survive, Heinrich's *Winter World* awakens the largely undiscovered mysteries by which nature sustains herself through winter's harsh, cruel exigencies.

**MIND OF THE RAVEN: INVESTIGATIONS AND ADVENTURES WITH WOLF-BIRDS**

In this book, Heinrich involves us in his quest to get inside the mind of the raven. But as animals can only be spied on by getting quite close, Heinrich adopts ravens, thereby becoming a "raven father," as well as observing them in their natural habitat. He studies their daily routines and, in the process, paints a vivid picture of the ravens' world. At the heart of this book is Heinrich's love and respect for these complex and engaging creatures, and through his keen observation and analysis, we become their intimates too.

Heinrich's passion for ravens has led him around the world in his research. *Mind of the Raven* follows a journey—from New England to Germany, and from Montana to Baffin Island in the high Arctic—offering dazzling accounts of how science works in the field, filtered through the eyes of a passionate observer of nature. Each new discovery and insight into raven behavior is thrilling to read, at once lyrical and scientific.

"Bernd Heinrich is one of the finest living examples of that strange hybrid: the science writer. . . . No definition of God has ever made me feel as comfortable, small, and important in the universe as Heinrich's insight into the mind of the raven."

—*Los Angeles Times Book Review*

**WHY WE RUN: A NATURAL HISTORY**

In *Why We Run*, Bernd Heinrich explores human evolution by examining the phenomenon of ultraendurance and makes surprising discoveries about the physical, spiritual—and primal—drive to win. At once lyrical and scientific, *Why We Run* shows Heinrich's signature blend of biology, anthropology, psychology, and philosophy, infused with his passion to discover how and why we can achieve superhuman abilities.

**Have You Read?** ***(continued)***

"A stunningly original book. It blends personal experience in world-class distance running with a firsthand account of the biology of running by one of its leading authorities." —E. O. Wilson

**THE GEESE OF BEAVER BOG**

With a scientist's training and a nature lover's boundless curiosity and enthusiasm, Bernd Heinrich set out to observe and understand the travails and triumphs of specific individual Canada geese living in the beaver bog adjacent to his rural Vermont home. Heated battles, mysterious nest raids, jealousy over a lover's inattention—all are recounted here in an engaging, anecdotal narrative that sheds light on how geese behave as they do.

Heinrich takes his readers through mud, icy waters, and overgrown sedge hummocks with deft insight, respectful modesty, and infectious good humor, accompanied by his beautiful four-color photographs and the author's trademark sketches.

"Heinrich's lyric writing and attentive observations make goose world come alive. . . . A pure joy." —*Los Angeles Times*